CAMPUS PUBLIC SAFETY
AND SECURITY

JAMES W. WENSYEL directed law enforcement, investigative and industrial security programs for the U.S. Army for more than twenty years. He subsequently served as Director of Security for several international corporations and continues to work as a security consultant. He has taught at the secondary school and college levels and developed security and public safety programs for a number of educational institutions.

CAMPUS PUBLIC SAFETY AND SECURITY

With Guidance as Well for High Schools and Private Secondary Schools

By

JAMES W. WENSYEL

CHARLES C THOMAS • PUBLISHER
Springfield • Illinois • U.S.A.

Published and Distributed Throughout the World by

CHARLES C THOMAS • PUBLISHER

2600 South First Street

Springfield, Illinois 62794-9265

© *1987 by* CHARLES C THOMAS • PUBLISHER

ISBN 0-398-05320-0

Library of Congress Catalog Card Number: 86-30204

With THOMAS BOOKS *careful attention is given to all details of manufacturing
and design. It is the Publisher's desire to present books that are satisfactory as to their
physical qualities and artistic possibilities and appropriate for their particular use.*
THOMAS BOOKS *will be true to those laws of quality that assure a good name
and good will.*

Printed in the United States of America
Q-R-3

Library of Congress Cataloging in Publication Data

Wensyel, James W.
    Campus public safety and security.

    Includes index.
    1. Universities and colleges--United States--Se-
curity measures. 2. Universities and colleges--United
States--Safety measures. I. Title.
LB2866.W46   1987   378'.196   86-30204
ISBN 0-398-05320-0

PREFACE

MOST SCHOOL administrators—from the President of a large university to the Principal of an urban Middle School—will admit concern for security and public safety at their schools.

Violent crimes, theft, drug abuse, arson, vandalism, and grafitti endanger the school community and campus facilities. Their effect is further felt in the growing inability of most Americans to meet the skyrocketing cost of education, much of which results from these violations of public law and order.

Most of these schools cannot afford professionally trained Security Directors or security staffs. The school's chief administrator, whether he has formal security training or not, remains responsible for protecting his school. He usually will delegate authority to a subordinate to run the campus security program. Unfortunately, that individual may have even less experience protecting a campus. This results in gaps in security coverage and can lead to on-the-spot decisions and make-do operations that are costly and of limited effectiveness.

This book suggests concepts of campus security and public safety and provides sufficient operational details to allow an individual who has limited security experience to develop and sustain an effective campus security program. It also will help administrators, teachers, and other school staff members, not normally acquainted with security matters, to develop better understandings of a school's overall protection needs and means by which they can be met.

This book then may be of help to education generalists and to those individuals specifically charged with protecting a school community and its campus.

INTRODUCTION

SECURITY and public safety are matters of genuine concern on every American campus. School administrators have a legal and ethical responsibility to provide a safe working environment for staff, faculty, students, and campus visitors. In addition, the cost of educating our youth has zoomed during the past decade. Much of that increase has come from similarly escalating costs of coping with crime and disorder on our campuses. If a school hopes to survive and to attract good students to sound educational programs, it must solve its campus security problems.

Budgetary restrictions limit the availability of professional security staffs. If the school's security program is to work, it must involve participation by the entire school community anyway. Their participation must be directed to get the most help from everyone involved.

This book suggests concepts in developing and maintaining good campus security programs. It can be very helpful to individuals already trained in security matters. More important, it can teach those who have little or no previous experience in general security matters, let alone in the unique requirements of campus security, but who find themselves responsible for their school's security and public safety programs. It suggests means by which staff, faculty, and students, as well as the nearby community, may be persuaded and led to properly support the school's security program. It suggests the studies that are necessary before a sound security program can be written and crisis management matters that should be addressed long before a crisis develops.

Solutions to security and public safety problems vary from school to school. The techniques cited in this book are valid in most schools. The reader, who finds himself responsible for protecting his school, should adapt them to his own situation.

If he does, and if he can develop a sound and workable campus security program, he'll have done his school and community a great service.

CONTENTS

CAMPUS PUBLIC SAFETY
AND SECURITY

CHAPTER 1

CRIME AND DISORDER ON CAMPUSES

STUDENT DISSENT of the 1960s, primarily over American involvement in the Vietnam War, was characterized by frequent and often violent campus demonstrations involving "sit-ins," "takeovers," destruction of college property, and angry confrontations among college administrators, students, and local police. All this was commonplace.

In the 1970s, as our involvement in Southeast Asia ended, so did this form of campus disorder. Unfortunately, it was replaced by an era of crime. Crime focused upon schools, from kindergarten to university. Crime against school property, crime against individuals associated with the schools, crime against nearby communities.

In June 1975, hearings before a Congressional subcommittee on elementary, secondary, and vocational education, heard testimony on the nature and magnitude of the growing problem of crime and disorder on school campuses.

At that hearing educators testified that during the period 1970-1973 they witnessed:

- 77% increase in assaults on teachers.
- 85% increase in assaults on students.
- 37% increase in robberies of students and teachers.
- 40% increase in rapes or attempted rapes.
- 53% increase in weapons confiscated on campuses.
- 37% increase in the use of drugs and alcohol on campuses.
- 11% increase in student dropouts, largely attributable to student fear.

Vandalism or theft of school property seemed to be as much a part of the scene as softball, the Model T, hot dogs and Coca Cola were to earlier generations. In 1974, the average annual cost per school district for vandalism was computed to be $63,000 ($135,297 in larger school dis-

3

tricts). The total cost for damage to school property in 1973 alone was reckoned at more than $500 M ($243 M burglaries, $109 M arson; $204 M other destructive acts).

Bare statistics, however, blur in reading. The total effect of school crime and disorder on the people represented in those statistics, and the cost in dollars and cents and labor and time lost, are hard to glean from rows of numbers alone.

To bring the problem into sharper focus, school administrators told the Congressional subcommittee:

Armed robberies are conducted in teachers' lounges, parking lots, residence halls, classrooms.

Students carry weapons of every description, despite the hazard of state laws against unregistered weapons.

School restrooms are taken over for gambling and narcotics; nonparticipating students are afraid to use them.

Gangs, often formed of local youths having no association with the school, brutally attack students, or extort "insurance" against attack. Motives range from personal gain to revenge for the victim's better grades or his/her conforming to school regulations.

Is the problem still not focused? Try these reported incidents:

In Texas, a teacher was raped three times in one semester. In one instance she was working alone in her classroom when the perpetrator, who it developed had no business being on school grounds, accosted her and, finding she had only $3 in her purse, forced her to disrobe then raped her.

In the Midwest, scores of pets at an elementary school were killed by vandals who broke into the school over Memorial Day weekend. More than 25 of the 40 classrooms were vandalized: plants uprooted, teachers' desks rifled, closets torn open and ransacked, records destroyed, windows and television sets broken, 30 tape recorders stolen.

In Florida, six youths, ranging in age from 6-11, set fire to an elementary school causing more than $200,000 damage. The boys also admitted setting a previous fire which caused more than $100,000 damage. None could provide a motive for either act.

Four fire bombs ignited in a science laboratory of an Eastern junior high school caused an estimated $260,000 worth of damage. Classrooms were destroyed; the school closed for many days. Two students, charged with arson, claimed they set the fire to get even with the principal.

In Dallas, a 15 year old boy was stabbed in the chest several times during a hallway scuffle; a shop instructor was clubbed with a mallet; a second student was stabbed by a 13 year old schoolmate after an argument on the school playground. All this in a single school day.

In California, a 17 year old honor student, about to enter the school cafeteria, was attacked and stabbed to death by six youths. None belonged to that school.

In a wealthy suburb of New York City, a bitterly contested high school basketball game was followed by a rock throwing melee as the visiting team attempted to load its bus. Before school authorities and local police could end the riot, one youth was killed and a number injured.

In Los Angeles, during an afternoon school dance, fighting broke out between rival student gangs. Weapons included steel chains, bricks, canes, sticks. Fighting spread throughout the school and into the nearby community. One death resulted.

Instances of vandalism of parked school buses (slashed tires, broken windows, cut distributor wires, sugar in gas tanks) occurred every day. In most instances, they caused school closings or delays in student transportation.

These instances occurred at the elementary or secondary school level. Crime and disorder, however, also struck at our community colleges, colleges, and universities.

Here the crimes included robbery, assault, petty thefts, and vast computer frauds in which students wreaked havoc in commercial data processing systems, sold forthcoming examinations, and were involved in large scale theft of books, computer equipment, and laboratory equipment.

Much of the crime and disorder on college campuses was violent. In January 1978, two Florida State University sorority sisters were raped and murdered in their university residence hall room.

Lone college students, studying in empty classrooms, were robbed, assaulted, or raped. Student suicides, attacks on staffs and faculties, and attacks on school property occurred often. Cults of all types and varied activities (often violent or disruptive) surfaced on campuses.

A late 1970s survey indicated that crime in public schools was far more serious than in the previous decade. During 1977 an estimated 157,000 cases of crime or disruption in public schools occurred in a typical month. Public school administrators estimated that nearly 15% of their teachers were hesitant to confront misbehaving students lest they too become victims.

In a survey of 375 colleges conducted during the early 1980s, 98% of the school administrators reporting stated that crime, particularly theft, was one of the greatest problems on their campuses. Associated problems included:

• General apathy toward the school and its problems.

- Difficulty convincing others of the need to professionalize handling of the school's security and public safety responsibilities.
- Faculty members refusing to face up to their responsibilities in security and public safety.
- Open campuses where individuals roamed at will and unchallenged about their identities or activities.
- Lack of key control throughout campus buildings and facilities.
- Lack of money to devote to a security and public safety program.

More and more school administrators are being forced to come to grips with the problem of crime and disorder on their campuses. Perhaps the most compelling reason is a genuine concern for the safety of their charges.

Several years ago a former Army officer became Commandant of a prestigious, nationally known military school. Security and public safety were among his responsibilities. The school was in dire financial straits. The campus was wide open. There was ample evidence of drug and alcohol use among students. Staff and faculty were apathetic, unwilling to accept disciplinary responsibilities. He was promised several additional supervisors to monitor the campus and residence halls.

Six months later he felt he'd done a good job. Barracks and students were much cleaner. The campus was restricted. Students and parents were responding to his efforts. Nevertheless, he resigned.

Citing inadequate funding, the school had failed to augment his security staff. Worse yet, it had accepted, at state expense, a number of streetwise, street-tough youths expelled from public schools. These youths prowled the barracks at night, bullying younger boys and selling "protection."

Late on a Saturday night, one of the gangs set upon a 13 year old boy alone in his room. They threw him on the floor and quickly piled on mattress, bed, locker and chairs. Had not one of the few Security Officers on random patrol come into the barracks, the boy would have suffocated.

That educator resigned because he could not guarantee the safety of his students. On a similar Saturday night, one year later, a 17 year old boy who had been drinking beer laced with PCP (an hallucinogenic known as "Angel Dust") stepped from an open third floor barracks window to prove that he could fly. He found, too late, that he could not. He died in the fall.

No school administrator would wish to carry the burden of that boy's death on his shoulders the rest of his life. For many, security and public safety on campus has a compelling personal urgency.

Another factor mandating genuine concern is the question of legal responsibility assumed by the school when a student enrolls.

That legal responsibility still has not been clearly defined. The older theory is that the school has a *locus parentis* role — that of an away-from-home parent who must, for the student's good, impose rules of behavior and punishment for non-compliance. That older concept changed somewhat with the Supreme Court's rule in Dixon *vs* Alabama Board of Education (1961) requiring stricter enforcement of the law for all citizens, regardless of age. "Ward" became "citizen," and the school was charged with enforcing public laws on its campus. That required the school to police itself.

The exact limits of the educator's authority and responsibility remain somewhat murky. Surely, however, he has a moral responsibility to provide a safe and secure environment for staff, faculty, students, everyone associated with his school's environment.

Additional impetus for enforcement of proper behavior include the skyrocketing cost in money and time lost for repairing or replacing damaged or stolen school or personal property and for school insurance premiums. Recently $8M was awarded a high school student permanently injured by a fall during his unauthorized (and unsupervised) use of school gymnastic equipment.

Invariably, the tax payer or the parent of a student is required to pick up the tab for all of it. And they're tired of doing that. These days one cannot look into a local newspaper without finding a headline: "School Bond Defeated," "School Budget Denied," "School Program Cut for Lack of Funds," "School Board Questioned about Deficit Funding."

This book suggests means by which the school administrator — whether at the kindergarten, primary, middle, secondary, college or university level — can determine whether his school has a security and public safety problem. It suggests problems he should be looking for and ways to resolve them. Whether the security department of his school consists of a large and well-trained, well-equipped professional security staff; whether it is a lone security "professional" supervising a contract guard agency hired to protect the campus; whether it is a faculty member wearing several hats, only one of which is security and safety; or whether it is a single man — principal of an elementary school trying to cope with security problems in his school building — this book can help.

Regardless of the size of a school's campus or of its staff and faculty, any successful campus security and public safety program must involve the school's staff, faculty, students, local police, and the local community in a joint effort.

Let us consider the general nature of that school community's involvement then look into the specifics of recruiting, training, and directing individuals from it in a school security program that really works.

CHAPTER 2

THE SCHOOL COMMUNITY

SOME YEARS AGO the movie "Blackboard Jungle" portrayed a talented and dedicated teacher attempting to cope with the disorder and violence endemic to an inner city school system. The movie suggested that had he been teaching in a suburban or rural school environment his disciplinary problems would have been almost nonexistent.

If that were the case in real life then, it does not apply today. The supposed calm of a suburban or rural school no longer exists.

Recently the son of a former candidate for Vice President of the United States was accused of selling drugs on a small Vermont college campus. So widespread were his alleged "services" that among other students he was known as "the pharmacist."

Drug and alcohol abuse, confrontations between students and teachers, and theft, misuse or destruction of school or private property are common occurrences in today's schools, public or private, urban or rural. The most prestigious school systems, sheltered in the most wealthy neighborhoods in American, have their own problems. Probably not so severe as in many public schools, but present.

There are obvious signs that a school has security and public safety problems. Such as instances of:

- assaults
- robbery
- extortion
- narcotics
- gambling
- theft of property
- misuse of property
- graffiti

- vandalism
- weapons possession
- drinking
- sex offenses
- trespassing
- arson
- bomb threats
- false alarms

Other indicators may not be as obvious but they too are valid signs of trouble. A walk across the campus, an hour in a classroom, or a chat with students or faculty will reveal them:

- Excessive absenteeism by faculty and students.
- Excessive cutting of classes.
- Frequent confrontations between students and teachers.
- Fear and tension among teachers (walking in groups, unwillingness to correct disruptive students, parking off campus to avoid reprisals, etc.).
- Hate literature on campus.
- Other signs of racial or ethnic unrest, such as a high assault or extortion rate among ethnic groups.
- Gangs of students clashing over "turf" or positions of prestige and power in the student body.
- Disturbances in classrooms, lunch rooms, cafeterias.
- Student demonstrations of disorderly character.
- Disturbances at dances, athletic events, and other student extracurricular activities.
- Unsafe areas (*e.g.,* rest rooms, library, parking lots) on campus.
- Frequent scuffling or fighting in crowded pedestrian areas.
- Unchallenged wandering of individuals, from within and without the school, throughout the campus.
- Uncorrected spread of rumors.
- Abusive or vulgar graffiti, particularly graffiti costly to repair or harmful to the school's image.
- Indifference by the staff, faculty, and students to graffiti or to other damage to school property.
- Student apathy toward school spirit, school prestige, etc.
- Reckless and abusive use of motor vehicles on campus.

A program to successfully control campus security problems cannot be developed overnight. The good administrator, realizing this, foresees problems and begins to build his defenses early, during a period of relative calm. If he does not, and if he must *respond* to situations rather than *prevent* them in the first place, his will be an uphill fight.

No security program will provide a "totally safe" environment. In a "totally safe" environment, nothing happens at all. In a school, things must happen. So the school administrator realistically must settle for an "acceptable risk." This book focuses upon that aim.

The campus security program must be continuous, not a series of starts and stops, emphasis and demphasis. It must be flexible and adaptable to changing situations. It must blend the school's educational and security functions. It must be characterized by caution, sensitivity, balanced judgment. It must involve the entire school community.

The school community: staff and faculty, students, local police, and the local community. These are the individuals who can make a campus security program work. The school administrator must enroll and judiciously use every element of that "community" if he is to adequately protect school property and protect individuals associated with that school.

His school may have use of many of the technical tools of the security trade: an excellent alarm system, good lighting, closed circuit television (CCTV) monitoring of sensitive areas, a response capability with a trained security staff. All will be of limited value, however, unless he also has the willing support and involvement of the human factor represented by teachers, students, school staff, and the local community.

I've suggested sure signs that a school has security and public safety problems. There are other signs, all related to this involvement by the school community, that suggest a *healthy* school.

First, is the school itself, an inanimate organism brought to life by the people who populate it, work in it, daily pass through or by it.

It should be the natural center for student and local community extracurricular activities. The more students and local residents use the school facilities, the less chance there will be for surreptitious theft, vandalism, or other harmful acts. The best defense against illegal acts is the presence of lighting, sound, active people.

Even if those people are not aware that they are a crime deterrent, and don't seek an active crime fighting role, their presence *is* a deterrent. And, if a person is involved in some way with a school, no matter how limited that involvement, it's difficult for him to resist some sense of responsibility toward that school. He will care what happens to it. If, for no other reason, than that he's using it; he needs it.

If the school presents a neat and clean appearance: painted buildings, neatly cut grass, pleasant landscaping, clear signage, well lit walkways and parking areas, streets and sidewalks free of litter, anyone near it (and certainly those involved in its operation) probably will feel some proprietary interest in keeping it that way. Built in security, if only in the form of "block watchers" to report illegal or harmful acts.

If the school has a reputation for dealing honestly, candidly, in a straight forward manner with everyone, it's going to engender a similar response. More built in security. This reputation is based on the school's administration: school board or boards of trustees; principal or president; staff and faculty. They're the ones who must be honest and fair dealing. They're the ones who must motivate the student body (the *heart* of the matter) to police itself and not to tolerate individuals or acts harmful to the school.

Idealistic? Certainly. Possible? Tough, but yes. Leadership? Of course.

School administrators must make the school, regardless of its size (university to primary school) seem small enough that the individual student does not feel lost or unimportant. Everyone associated with the school must be made to feel that his/her own personal contribution can cause an inanimate block of buildings, strung together by walks and roads, to come alive and assume a personality and character of its own. The concept is called "leadership." It will be suggested throughout this text.

Incidentally, throughout the text I usually will use the pronoun "he" to avoid the cumbersome "he/she" assurance that I am not biased regarding the role of women in leadership positions on the campus. They surely have their role and contribute every bit as much as their male counterparts. To constantly use "he/she," however, detracts, from the flow of the text so I will avoid it. I trust that he/she reading this book will understand and bear with me.

Similarly, I may refer to the individual formally responsible for campus security and public safety simply as "the Security Director" (and to his overall responsibilities simply as "security"). Whatever his title, his responsibility is the same. And it's a large order.

The staff and faculty must understand that the student body is the heart of the school community. The students are the school's reason for being. The students offer its best hope for survival and health. Teachers and administrators must know, respect, and use formally established student leadership chains (student councils, class officers, athletic captains, club officers, etc.). At the same time, they must realize that for every "elected" student leader there are many others who are *natural* student leaders. These natural leaders may not hold a student office, but they have their own following and their own ideas. They want to be heard and they should be heard.

Terribly important, the staff and faculty must *know* what is happening among their students. They should know student leaders; they should know student concerns. They must establish and maintain dialogue with their charges. It does not come with their titles, nor with the authority they wield in the school's administration. They get it, as the television ad suggests, "The old fashioned way." They earn it.

If they've done that, if they have the students' trust (perhaps not always their affection but always their respect and trust), they are not going to be very surprised when a problem surfaces on campus. They'll have had some indication of it beforehand.

Somewhere, sometime — in a classroom, over a cup of coffee with a group of students discussing modern poetry, during rap sessions on the ball fields, in a corner of the gym at the junior prom — some student will share a concern or a confidence. That sharing may lead that teacher or administrator to realize a problem is developing at his school.

Leadership. Leadership by caring, sensitive, mature, intelligent adults who are more concerned with the well being of their charges than with their own popularity. Who care more about getting the job done than about making the job easier. The student first. Always. Sooner or later that student will recognize which teacher, which coach, which Security Officer really is concerned about him and about the school. Which can be trusted. And he'll confide in him. This book mentions "leadership" a lot. It's the best security device ever developed.

The heart of a healthy school is a healthy student body. Here are signs to look for:

- Formally defined and established student leadership working with school administrators in procedures to help govern and develop the school. The procedures work; students know they work.
- Informal leaders and "outside groups," however, also have established ways to make their views known. These ways work; students know that too.
- Most students are concerned with the school's image.
- Most students are concerned about law and order on campus.
- Most students work together in a school crisis (a building burns and school property must be saved; hate literature surfaces on campus and is refuted by the student body, etc.).
- Many students volunteer to help worthwhile school projects.
- The students, not the Admission Director, are the school's best recruiters. .
- Students understand and accept their roles, rights, and responsibilities as part of the school community.
- Students know what is happening among themselves and often will police themselves (*e.g.,* care of residence halls, ousting the drug pushers, etc.) on their own initiative.

The local community plays its role too by participating in school activities, by opening its doors to students, by encouraging town officials to cooperate with school authorities, by showing as much concern about crime on the campus as they do about crime on their city's streets.

Local police can do much to enhance a sound campus security program. If they work well with the school's Security Director, there will be a continuing exchange of information about potential or existing law enforcement problems that affect both school and community. There also will be physical support or back-up by local police *when* the school administrator wants it and, very important, the *way* he wants it.

Later, we will consider the roles of campus police and town police in securing the school. Each is important. Each should support the other. Local police must do all they can to avoid students' adopting a "town cops," "campus cops" differentiation, with all that implies.

The school's security staff, whether it be one man, contract guards, or a large proprietary guard force led by a security professional, must walk the delicate line of enforcing public law and school regulations on campus, on the one hand, and of being the school's walking public relations department on the other. Security Officers and campus investigators must *earn* and *keep* the respect of staff and faculty and students. If they do their jobs well, they too will recognize developing problems before those problems become critical. If they do, as often as not, they'll be able to keep the lid on those problems, or resolve them, and no one will even know they did it. That's part of their job. More on this later too.

For any campus security program to work, the willing involvement of the school community is critical. Once won, it must be channeled in the right direction and kept moving in that direction.

Key to that is developing an awareness of the need for a good campus security program. That means education. Education: a reasonable place to start. After all, we're talking about a school.

CHAPTER 3

SECURITY AWARENESS PROGRAMS

IF YOU WANT someone's help in an important project, you must convince him that the project is worthwhile and that you really need his help to complete it.

Americans are reluctant to become involved in security programs. They generally are law abiding themselves and prefer to leave those who are not to someone else's care. Ask the faculty's help in a school's security program and often the response will be, "Hire a cop" (meaning a local Burns, Pinkerton, or Wells Fargo security guard), or "Let George do it. He's in uniform," or "He's being paid to do it" (meaning let the staff or faculty member charged with running the school's security program handle it single-handed).

If you are the school's administrator, experience will tell you that neither answer will work. "Hiring a cop" or leaning hard on the one man charged with the school's security, without involving the whole school community in the program, simply won't get the job done.

The answer is to persuade as much of that school community as possible that there is a security problem at the school and that each individual can and should play an important part in handling it.

That persuasion will not be easy. Worthwhile things seldom are. Some individuals may never contribute or, worse yet, may actually hinder the program. If key figures in the school, however, use dynamic, enthusiastic, and knowledgeable leadership in approaching others, most of the staff and faculty and student body can be won over. That will be enough.

The "winning over" begins with the initial orientation given the new administrator, teacher, worker, students, security staff member, local police chief, parent or nearby resident involved on a continuing basis with the school.

It is reinforced by informal "rap" sessions and periodic briefings or special orientations (as at the start of a school year or before a critical school event); by good use of pamphlets, films, posters publicizing the security program; and by operation of special security services (escort services, school crisis center, etc.). However the school's security program is publicized, its purpose is to convince every person in every element of that school that his help is necessary if the school is to protect its property and insure a safe and secure working environment.

Whenever possible, security and safety orientations should be introduced by the head of the school or his official representative so that his support of the program is clear.

A lead role by the individual specifically responsible for the security program (*e.g.*, Security Director, Vice President for Administration, Assistant Principal, etc.) also is critical. Responsibility for important security orientations, for example, should not be handed, as a matter of course, to an available shift supervisor of the campus police department.

Educators are prone to critique another teacher's presentation. They're also prone to focus so intently on the manner of presentation that they miss the message itself. All security briefings should be given by an individual who clearly knows his subject and is at ease with it. One who can hold his own in any exchange with any other faculty member (for he truly is a faculty member himself).

Normally that means the head of the security department or a very carefully selected representative. The participation of the Security Director conveys official emphasis that somehow is missing if an assistant or, worse yet, an ex-cop turned Security Officer presents a rambling, disorganized, replete with factual and grammatical errors presentation to a skeptical and not too receptive audience well sprinkled with masters and doctoral degrees.

In the long run, however, on an informal basis the best representative of the school's security program is the individual Security Officer. The professionalism, bearing, enthusiasm, knowledgeability, and interest he displays as he walks his daily patrol and casually chats with students and teachers go a long way to persuade everyone of the worth of the program he represents. That's why the selection, training, and management of the security staff are so important.

Whenever possible, let the head of the security department make the initial overture to members of the school community. This should be followed by other means of publicizing the school's security program, including the use of individual Security Officers.

The following suggestions for the educational aspect of a campus security program are designed for a large college or university. They are adaptable, however, to any school of any size and thus merit consideration by whomever is responsible for campus security.

Security, in any organization, is only as strong as the administrator of each department wants it to be. So he or she is the individual we must first recruit to the campus security program.

The initial orientation of school department heads might include the following topics:

- Extent and nature of security problems on campus; particular security problems.
- What the security department is doing to try to solve them.
- Formal school organization and its chain of command.
- Student government organizations; leaders; procedures.
- Informal student leadership (organizations, individuals, means by which they make their views known).
- Organization, functions, general operational procedures of Security Department.
- Responsibilities, role, degree of school coordination with local law enforcement agencies.
- Communication for emergencies: police, fire, medical.
- Presentation of written copy of school rules and regulations, student orientation pamphlets, etc., with the strong injunction that they read them.
- Legal and moral security responsibilities of administrators to their staffs; leadership roles expected of administrators.

Subsequent orientations should reinforce security procedures and concepts and keep administrators current on campus organizations, leaders, and happenings.

Initial orientations for faculty members should contain all the information suggested for department heads, plus the following specific suggestions:

- Keep your department head informed on all matters affecting the security and safety of the school.
- When appropriate, recommend better ways to do a job.
- Support the school's security and safety program, the security department that operates it, and the local law enforcement agencies which reinforce it.
- Get involved with your students, inside and outside the classroom; cultivate their trust; be alert to security and safety problems.

- Report security or safety hazards or violations.
- Report criminal violations; no exceptions. Be willing to testify about them.
- Know the signs of drug abuse and how to handle it.
- Be concerned about the prestige and public image of the school.
- During class changes or other crowded student activities, be in hallways or where the teacher can see, and be seen, and be heard; take corrective actions.
- Support less experienced teachers.
- Turn in attendance and other required reports promptly; be sure they are accurate.
- Be willing to meet with parents and to tell them of student tardiness, absences, misbehavior or questionable conduct.
- Challenge individuals who are not known to you and who appear lost or confused, are in unauthorized school areas, or are engaged in questionable conduct.

Key to the success of any campus security program is faculty involvement: involvement with security, involvement with students. This involvement will help the Security Director foresee and avoid or minimize the effects of campus security problems.

Another group critical to the Security Director is the support staff: secretaries, teachers' aides, carpenters, painters, nurses, drivers, housekeepers, groundskeepers, athletic trainers, etc. As they work and move about the campus, they talk with students, develop sources of information, see a great deal of everyday life that probably is missed by department heads or teachers.

Their orientation, in addition to many of the things suggested for administrators and faculty, must assure them that their help is critical to the school's security and safety. They must be persuaded to become personally involved with the school. Simply putting in their work hours and leaving for the day without concern for the school that provides their livelihood, just won't do.

They should report security and safety hazards or violations, problems among faculty and students, troublemakers observed on campus. They should question the identity and purpose of any suspicious individuals on campus. The query can be tactful and courteous, but the very asking will cause that individual to realize that he has been noticed and will be remembered. If he means to do the school harm, that realization may cause him to hesitate.

Formal student orientations should emphasize their responsibility as young adults to keep this portion of their society safe and a pleasant place to work. Pride in themselves and in their school must be emphasized from the start. The person giving the orientation must demonstrate an appreciation that most people are law abiding and considerate of others. From the context of his talk and the way he presents it (and himself), his audience will be persuaded that Security Officers will cooperate fully with those individuals and devote their "enforcement" activity to the few who are not law abiding.

Other topics that should be stressed in student briefings include the following:

- Extent and nature of security problems on campus; type crimes; areas that are particularly vulnerable.
- What the Security Department is doing to try to solve those security problems: alarms, lighting, locks, etc.; mobile and foot patrols; campus investigations; cooperation with local police.
- Formal school organization and its chain of command.
- Student government associations; leaders; procedures.
- Mission and functions of the Security Department.
- Presentation of a security pamphlet citing school rules and regulations, traffic procedures, means to secure personal property, safety, lost and found procedures, other security matters.
- Emergency communications: police, fire, medical.
- Matters for specific emphasis: drugs, vandalism, graffiti, bicycle thefts, etc.
- Availability of counselors, representatives of the Security Department, special security services (escorts, property identification marking, etc.).

Informal "rap" sessions between teachers or security staff members and students are very important follow ups to these initial orientations. They develop trust and reinforce the security and safety messages. They also may surface student concerns.

Similar informal sessions between the school's staff and the local community are very important. At open houses, neighborhood appearances, or liaison visits between school and community officials, visitors or individuals residing near the school should be encouraged to cooperate with school and local police departments.

The school community must be reminded constantly of concern for crime and disorder on campus. There are many ways to do this in a low-keyed but effective manner that will not alienate anyone. Approaches are limited only by the imagination of the school's administration and students enrolled in the campus security program.

The danger of crime, disorder, or safety hazards on campus might be emphasized and reinforced by timely, provocative, effective posters and pamphlets; by advertisements or cartoons in the campus newspaper; by spot announcements on campus radio or television; or by campus comic books. The latter might depict the school mascot, for example, in a crime fighting or accident fighting role. Posters might include humorous cartoons and poetry carrying a security message. The uniqueness of its presentation may cause its message to be remembered.

Security Officers, noting improperly parked or unlocked vehicles or bicycles, unlocked residence halls or sensitive laboratory areas, student property left unprotected, fire hazards in dormitories, or other security or safety hazards might leave pre-printed cards annotated to show the specific areas of concern and politely calling it to the attention of the responsible individual.

Done properly, this approach engenders good will between the recipient and the school's security staff. After all, anyone would rather be politely corrected than be formally cited for non-compliance with a school rule. Citations often are necessary, of course, but suggestions and warnings have their places too and often are very effective. The basic aim is not to apprehend a rule breaker but to avoid the rule's being broken in the first place or, if it is broken, to minimize its effects.

Most major colleges and universities have security education programs of the type suggested here. The individual responsible for his school's security program normally need look no further than the nearest large university for appropriate pamphlets, posters, and other devices that urge individuals to accept responsibilities toward campus security problems.

Formal and primary responsibility for campus security, however, remains with the school's Security Department. We turn now to that department's responsibilities, composition, and functions; then to selection and training of its members.

CHAPTER 4

CAMPUS SECURITY DEPARTMENT

THE SCHOOL'S security department, however it is constituted and by whatever name it is known, has the following responsibilities:

- Protect the lives and safety of all individuals within the school environs.
- Safeguard school and personal property from theft, damage, vandalism, or misuse.
- Prevent or control crime on campus.
- Investigate crime on campus, apprehending offenders and recovering stolen property.
- Preserve peace and resolve conflicts on campus.
- Provide an immediate response to emergencies and to all persons requiring aid on campus.
- Enforce school policies, rules, and procedures.
- Advance cooperative relationships within the school community and contribute to the school's public relations program.

It accomplishes these responsibilities by using Security Officers and other members of the security staff who are professionally selected, professionally trained, and professionally directed in their operations; by granting students real rights and real obligations of school citizenship; and by involving all members of the school community (particularly the student body) in the campus security and safety program.

Its responsibilities normally are limited to the school boundaries and to individuals working in or passing through the school environment, but these responsibilities are very broad and, in one way or another, touch upon every facet of the school's life.

School administrators may approach this security and safety responsibility in several ways:

- Do nothing; have no formally developed security and safety plan or department but rely upon the staff and faculty for enforcement of school and community laws for maintaining order and protecting property.
- Employ local police officers in a "moonlighting" security role or as an officially sanctioned extension of the local community's police jurisdiction.
- Hire contract guards, supervised either by the school staff or by the contract guard agency.
- Employ a proprietary security staff under the supervision of a school administrator.
- Employ a combination of the last three techniques.

Experience has proven that to do nothing, or to rely solely on a school's staff and faculty alone for campus security, usually will not work.

Voluntary compliance with established rules for law and order or individuals' harmoniously working with others (often to one's own detriment, as in motor vehicle traffic) are just too much to expect. And attempts at enforcement by staff and faculty members alone usually require technical skills they do not possess and detract excessively from their primary duties.

Employing local police, either as "watchmen" or moonlighting campus patrol officers, or placing the campus under the town's law enforcement authority, also have not proved satisfactory.

Law enforcement on campuses differs from that in towns. The student with whom campus police must deal differs from his town counterpart. So is the school administrator's intent in meeting this portion of his responsibilities. His concern is cheerful acceptance of his school's rules (or, at worst, non-compliance held to a small roar), not with the apprehension of law breakers. Town police see student transgressions differently than would the staff and faculty. Similar differences in the approach of a campus Security Officer and a town police officer, as well as problems of availability when needed, loyalty, and conflicts of interests all make it better to keep town law enforcement and campus law enforcement cooperative and mutually supporting but entirely separated in their execution.

If a system using town police to patrol the campus is considered, however, the following matters should be formalized in a memo of understanding between the school's administrator and the community's police department long before a town police officer begins his first patrol on campus:

- Number of police officers to be assigned to campus duty.
- Police officers' deployment and use on campus.
- Whether the same officers will be assigned to the school every day and whether the school will be the *primary* assignment for those officers.
- Whether assigned officers will be "on duty" officers or "off duty" officers on special assignment.
- The school's administrator is the final authority on selection, assignment, and retention of police officers on campus.
- Only the school's administrator can request extra police officers on campus.
- The school administrator remains in charge of the overall campus security and law enforcement plan with the appropriate town official, under him, having operational control of police functions on campus.
- The school administrator can mandate any special training to be given assigned officers (*e.g.,* handling juvenile offenders), the amount of training, where it will be conducted, and under whose responsibility.
- Pay, uniform, equipment (including weapons) of police officers on campus.
- When requested, police services will include the school's extracurricular activities.
- Jurisdiction and method of handling complaints.
- Policies and procedures for arrests on campus.
- Student records remain confidential and under control of the school administrator.

Many schools use contract agency security guards for campus security. There are advantages. They cost the school much less (less base salary, no benefits, no medical compensation, no paid absences, no uniform costs, no equipment costs, etc.). The school has no training responsibility and can dismiss a Security Officer or hire more by a simple phone call to the contracting agency. Very important, the school *may* not be required to provide insurance coverage for contract guard operations on campus or for contract guard actions resulting in legal suits. Finally, the school is not burdened with administrative or operational matters related to the security force as these normally are handled by the contract guard agency.

The disadvantages to using a contract guard force, however, are substantial. They include a frequent turnover of personnel, personnel who have been poorly selected and poorly (if at all) trained. These individuals usually are paid marginal salaries, offered few if any benefits, and respond in kind. Most have not been given a background investigation (turnover in the guard agency is so high that the cost of a background investigation would be prohibitive). In sum, unless they have really fine supervision, their motivation, loyalty, dependability, reliability, and overall job performance are apt to be marginal. Unfortunately, contract guard supervisors usually are not fine supervisors.

Contract guard agencies, seeking a lucrative contract, promise much. Their early performance usually is good, certainly better than the agency they replaced. Most, however, begin to lose their incentive almost as soon as the contract is won. Within a few months the school administrator may notice excessive turnover in assigned Security Officers, continuing problems with vacant security posts or with punctuality and attendance. Most critical, the excellent Security Officers he met during the agency's contract presentation, or saw during the first week of the new agency's patrol of his campus, suddenly disappear. After he's asked a certain number of questions about that phenomena (and finally demanded an answer), he learns that they have been transferred to a new and/or more lucrative account with a less qualified replacement guarding his school. He'll also find increasing evidence of sloppy bookkeeping by the contract guard agency: charges for guards who did not work, charges for hours not accounted for, sudden escalation in overtime charges (a guard held over for a second shift because his replacement had not appeared, etc.). He may also note increasingly poor appearances by many guards as the quality of their supervision diminishes. At the same time, there will be a rash of confrontations between the contract guards and the rest of the school community. In short, he suddenly has many problems and, within a few months, will be looking for another contract guard company.

If he elects to stay with the contract guard concept of campus security, he can ease many of his problems by assigning school employees to supervise the contract agency's services. Contract guards, for example, would have school employees for their shift supervisors.

Larger or better endowed school systems often have their proprietary security staffs: a security and public safety department comprised of school employees. If a proprietary staff costs more, it also has better selection, training, and operational control of personnel who are better motivated than their contracted contemporary and more apt to do a good job. Individual Security Officers are going to respond better to the higher wages and other benefits offered by the school, particularly if service in the security department can be linked to a school-wide career development program.

Objections to a proprietary staff include the much greater cost in salary and benefits and the greater time and effort spent administering the overall program (uniforms, sick benefits, liability insurance, training, career development, etc.).

The best solution, other factors (personnel available, financing, etc.) permitting, is a proprietary staff or a mix of proprietary and contract guards with proprietary supervisors and proprietary Security Officers in more critical campus areas, contracted guards in less critical areas or mixed in with the proprietary staff (as in campus special events requiring additional coverage).

Just as there are different approaches to staffing a campus protective force, there also are different views of where the Security Department should fit in the school's organizational hierarchy.

Sometimes the department is placed under the control of the Director of Physical Plant, sometimes under the Director of Administration, sometimes under the Dean. More rarely, it functions under the direct supervision of the chief school administrator (president, principal, etc.). In any of these positions there may be problems, but some present more than others.

Most often it has been placed within the Department of Physical Plant or the Department of Roads and Grounds. This is the department responsible for roads and grounds, buildings, environmental controls, maintenance of school property and equipment, and similar logistical matters.

If the Security Department operates under such a logistical department, its members probably will be used to schedule repairs, maintenance and similar functions, or to respond to a teacher's complaint about a leaky faucet or a long overdue work order for a new blackboard. That should never be. The department is there to protect the school. It must not be diverted into other matters. Another difficulty is that the Security Director would be working for several bosses, just as does the Director of Physical Plant. Serving several masters is difficult at best. When it is concerned with such sensitive matters as are commonplace in campus security, the problem is heightened many times over. A final difficulty is that, if the department is physically located with the physical plant offices, it is going to be in the basement of a building apart from the school's mainstream. Not good for security and public safety, as discussed below.

Some of these problems also occur if the department is placed under the Director of Administration, but they are somewhat less as the mutual areas of concern for the two departments are both people oriented and more closely linked. Key to this arrangement would be the authority wielded by the Director of Administration, the time he is able to devote to the Security Department, his trust in his Security Director, and his accessability to the school's head for the support a good security program needs.

Security and public safety functions should not be assigned to the Dean. He has too narrow a perspective. His interest and character normally are in the functioning of the academic departments and have nothing to do with the many other facets of school operations (roads and grounds, residence halls, sales facilities, fire and medical protection, etc.). In addition, the Dean often is not high enough in the decision making hierarchy to adequately support security operations. He also would be miscast as his role primarily is one of an adjudicator, not an enforcer. Counseling is his stock in trade, not assuring public safety and order or, when required, assuming a firm law enforcement stance with the school community.

The Security Director usually would prefer to work directly for the chief school administrator. That individual has the clout that Security Director needs, is more free of political infighting, can cut across all departments of the school, can make necessary decisions and provide necessary support. His office has the high visibility and authority needed for a good campus security program. The problem, however, is that the head of a school is a very busy man and often simply would not be able to devote adequate time to security matters, perhaps even to assure a listening ear to the Security Director's problems.

The best solution varies with the school and with the personalities involved, but key to it is that the man responsible for security should be high enough in the school's hierarchy to assure support for his program yet be accessible to the individual (Security Director) who will administer it for him.

Linked to this suggestion, the location of the campus Security Department should be highly visible, of impressive professional appearance, and easily accessible to everyone having proper business with it but capable of being defended against demonstrations or more violent disorders.

The Security Department should have its own budget, primarily developed by its Director, and providing for adequate staffing with reasonable pay and benefits for its members, adequate equipment, adequate funding for training and public relations programs. If all those "adequates" are not possible, it should still be *his* budget so that he clearly knows the resources available to his program and his responsibility to use them in the way he feels best accomplishes his department's objectives. He should not have to go, hat in hand, to one or more department heads, to justify, on a case by case basis, every expenditure he feels necessary. Let him run it, under the overall direction of the school treasurer or whoever has fiscal oversight responsibilities.

The size of the Security Director's staff depends upon such things as:

- Size of the school to be protected.
- Size and type campus (urban/rural, compact/vast, modern buildings/old physical plant, etc.).
- Student body (size, type student, day school/boarding school).
- Degree of protection required (sensitive areas, etc).
- Technical equipment (alarms, CCTV) available.
- Local police involvement on campus.
- Quality of security department's supervision.

The campus Security Department might be functionally organized as shown in Figure 1. The suggested organization reflects a large staff handling a complex campus security problem and can be modified as required. Each functional area need not necessarily be the responsibility of a different individual. It would be ideal if that much attention could be given to each area but, budgetary constraints being as they are, it may be necessary to assign several hats (thus several responsibilities) to the same individual. For example, all fiscal responsibilities might be handled by one person, rather than by three or four individuals; the individual responsible for security surveys also logically might handle the campus crime prevention program, etc.

Within the major functional areas suggested, specific responsibilities might be assigned as follows:

a. Security Director
 - Directs the campus security and public safety program.
 - Responsible for overall administrative, operational, and fiscal matters related to his department.

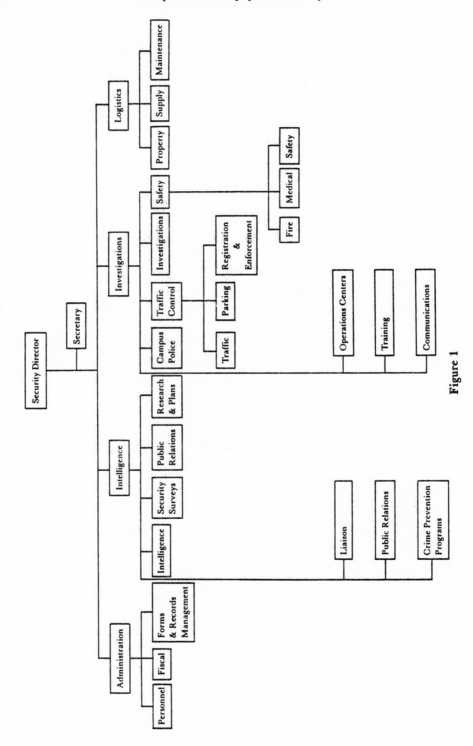

Figure 1

b. Administrative branch
  • Responsible for all the department's personnel functions.
  • Responsible for department's budget, purchasing and contracting.
  • When requested, can conduct informal audits of school's revenue producing or purchasing and contracting operations.
  • Maintains the department's administrative and personnel records.
c. Intelligence branch
  • In coordination with the Operations branch, conducts information evaluation and collection programs.
  • Evaluates the department's operational policies and procedures, equipment; recommends changes.
  • Conducts research on campus security problems and recommends changes to security procedures.
  • Develops special operations plans and Standing Operating Procedures (SOP), *e.g.,* bomb threats, natural disasters, handling large crowds at special events.
  • Conducts security surveys, fire prevention surveys, inspections and spot checks; recommends measures to correct noted discrepancies.
  • Responsible for the department's public relations activities.
  • Responsible for school's crime prevention (orientations, brochures, film spots, training programs, marking of school and private property, etc.).
  • Conducts liaison with local law enforcement and emergency agencies, other schools, etc.
  • Maintains all records of Intelligence branch.
d. Operations branch
  • Responsible for all police functions (patrols, alarm monitors, crowd control, campus traffic and parking, vehicle registration, etc.).
  • Responsible for campus safety program.
  • Conducts covert and overt investigations of criminal matters, violations of public and school laws or regulations, personnel background inquiries, accidents.
  • Responsible for campus emergency operations center operations.
  • Responsible for campus radio communications.
  • Responsible for Security Department's overall training program.
  • Maintains all records of Operations branch.

e. Logistics branch
  - Assures proper accountability for all property, supplies and services assigned to the department.
  - Maintains and issues all departmental supplies, assures availability of emergency or standby supplies and equipment.
  - Responsible for proper maintenance of all departmental equipment and facilities.
  - Maintains all logistic records related to the department.

Individuals selected to head each of these functional branches must be technically well qualified to work in those specialties. They should be flexible and adaptable in their thinking, young in spirit (if not in age), and enthusiastic about their work. They will put in long hours and be subject to call at any time. It's the nature of the security beast and, if they cannot understand that and cheerfully work under its constrictions, they should work elsewhere.

They should be good administrators who are capable of thinking on their feet and expressing themselves well (orally and in writing). Occasionally there are exceptions to this last item on the security administrator's list of attributes, but they are few. I knew an Army Motor Sergeant who, soon after reveille each morning, posted a list of "Work to be Did Today." In his position, working with vehicle mechanics whose educational level averaged less than a high school diploma, his grammatical errors were overshadowed by his technical skills, enthusiasm, personal leadership. He probably would not have survived confrontations with a school's staff and faculty, however, as they'd likely be stopped by the "Work to be Did" and never see the work that was done. Branch chiefs should have at least a high school education and be able to express themselves clearly.

The Security Department's staff, from top to bottom, should be emotionally stable. Frustrated, would-be cops have no place in a campus security force. The job involves a lot of personal stress. Confrontations between an excitable campus Security Officer and staff, faculty, or students are the last thing a Security Director needs.

All members of the department should be encouraged to be warm, friendly, courteous, very professional in demeanor and in their job performance. They should be discrete in all their actions, professional and personal, on and off the job. Particularly in their speech. They often are privy to sensitive and confidential information regarding campus figures and must resist the temptation to share it with others.

All security staff members must be neat and trim, presenting a good personal appearance. They must be physically fit. Every man, whether he normally works behind a desk or patrols the campus, must be physically capable of running, jumping, climbing, doing things he might have to do in an emergency.

I speak of a military approach to all aspects of security operations. If the Security Director believes in these attributes, personally practices them, and insists that everyone in his department do the same, it can be done.

Previous work experience for staff members should be people related. That is the name of the game in campus security. Formal police training and police experience can instill habits and concepts that do not apply to campus security (*e.g.*, emphasis on lawbreakers and on apprehension rather than on prevention). Supervising individuals who have a police background may be more difficult as they often are less flexible, less adaptable, more apt to challenge their supervisors. The ideal Security Officer or supervisor might be an individual who has a high school education, is interested in earning a college degree, and earned noncommissioned officer rank in one of the military services.

All security personnel should undergo a thorough background investigation, hopefully conducted by department investigators. It should verify information (education, work experience) the applicant has provided; check all pertinent law enforcement records; verify the applicant's good character, emotional stability, and personal habits.

Key to the success or failure of the Security Department is its Director. By whatever title he is known (we'll normally call him the Security Director), his professional competence, enthusiasm about his job, and personal attributes will set the tone for his entire staff. If he is an enthusiastic, hard-working, highly conscientious individual, his staff will reflect it.

He should have a well-rounded background as an administrator at departmental level and several years' practical experience in security, safety, and law enforcement. That experience need not necessarily be related to a campus. If he already has a solid base in general law enforcement, security, and safety, he can learn the additional aspects of those fields that are unique to a campus environment.

I do not advocate the practice of hiring an individual to protect a campus solely because he is a former agent of the FBI or CIA, or an ex-police chief. Those credentials alone are not enough. Being an Indian does not make one a Sioux warrior. Being with the FBI or CIA does not mean that the individual has experience in more than one or two phases of general security. An applicant may have spent twenty years in "federal law enforcement" or "national security" but all twenty may have been restricted to investigating one or two type crimes or working as an analyst with very limited supervisory experience. More often, that individual also will have had very limited experience with such things as alarms, supervising a guard force, security surveys and inspections, traffic control, or preparing detailed operational plans and procedures.

As with all "flat out rules," of course, there are exceptions. But I do suggest that the school head look beyond the titles of his applicants' resumes to their specific job experiences before settling on the man to be responsible for the security and safety of his school.

His Security Director also will be the primary public relations representative for that department — a very sensitive department that should be involved with every facet of his school's operations. The individual should be at ease before all types of groups, particularly so with students, and capable of representing the school well before the public.

The school has a legal and moral responsibility for due care in hiring and training its employees, and for their job assignments and supervision. The earliest place to catch a potentially unsatisfactory employee is in the hiring process itself. If he is never employed, he cannot harm the school or cause it embarrassment. Hiring standards and employment screening are very important.

The Security Director should establish and enforce strict employment standards. They should be written and should be displayed in the department's personnel office. They must not discriminate in terms of sex, race, or creed, but they might include the following requirements:

- Applicant must be able to adequately read, write, and speak the English language.
- If applicant is to be armed, he must meet local, state and federal licensing requirements and satisfactorily complete a prescribed course of instruction in weapons safety and in the use of the specific weapon he will carry.
- Applicant has no previous felony convictions.
- If applicant must operate a motor vehicle, he must possess or be capable of obtaining a valid motor vehicle operator's license and have a good driving record.
- Applicant is not addicted to drugs and has no record of drug or alcohol abuse.

- Applicant is in good physical and mental condition as determined by the school physician's examination and has passed a physical proficiency test.
- Applicant has a good work record (satisfactory work habits and free of dismissals or other punitive actions by previous employers).

Training of the security staff should develop technical competency, confidence, dependability, and reliability, courtesy, discretion.

It should consist of an initial training period successfully completed before the individual is assigned operational duties. The initial training must be followed-up, as often and as thoroughly as operational requirements permit, with extensive additional training (formal, informal, on-the-job correspondence, etc.) to develop technical competency, improved job attitudes, professional confidence and demeanor.

Tuition assistance and genuine encouragement for all staff members to participate in college continuing education programs in security or a related field also are very beneficial to the school and well worth their cost. Some will find that this education leads them to other employment; some will move from the security staff to another part of the school's administration. Those risks are worth taking in light of the overall value of employee development.

Whenever possible, Security Officers should be trained by their own security staff. The Security Director has better control of lesson content, the quality of instruction, student motivation and participation. He can see students at work, first-hand, rather than receive a third-hand report of their progress. Thus he can evaluate programs and students for himself.

If Security Officer training is done by contract with the local police department, it may cause confusion among campus security trainees over the concepts of law enforcement *vs* campus assistance, or arrest *vs* containment of criminal acts. Finally, training in a police academy often results in the best students' being recruited by the police department.

Training methods might include lectures, seminars, film clips, movies or slides, tape cassettes (many commercial ones are of good quality and not too expensive). All classes should feature maximum student participation (discussion, questions and answers, practical exercises, tests).

Appropriate training subjects might include:

- Patrol techniques (mobile/foot)
- Detex systems (if used on campus).
- Communications procedures and codes.

- Investigations.
- Interviews.
- Interrogations.
- Legal rights/obligations/pitfalls of Security Officers.
- Legal rights of private citizens.
- Criminal law.
- Pertinent state and local laws.
- Working with local law enforcement agencies.
- Apprehension of suspects.
- Arrest of suspects (legal pitfalls, arrest procedures).
- Report writing.
- Department reports.
- Protecting crime scenes.
- Rules of evidence.
- Handling crowds.
- Traffic control; traffic enforcement.
- Search and seizure (legal requirements, procedures, chain of evidence).
- Emergency medical measures.
- Fire prevention; fire fighting.
- Handling juvenile offenders.
- Assisting physically handicapped.
- Handling drug abuse, mentally disturbed individuals.
- Personal attributes of Security Officers (appearance, discretion, bearing, demeanor, physical fitness, job performance, etc.).
- School facilities
  - Roads and grounds.
  - Operational use of each facility, within each facility.
  - Sensitive security and public safety areas.
  - High crime areas.
  - Physical layout of facilities: windows, doors, exits, electrical and water controls, lights, HVAC controls, alarms, CCTV coverage, fire equipment, etc.).
- School rules
  - Traffic.
  - Vehicle registration.
  - Parking.
  - Student restrictions.
  - Restricted areas.

- Department rules
  - Personnel procedures; benefits.
  - Attendance; punctuality.
  - Rules of conduct.
  - Appearance; uniforms (allowance, maintenance, proper wear).
  - Discipline.
  - Reporting procedures.
- Emergency procedures
  - Fire.
  - Bomb.
  - Natural disaster.
  - Alarm response.
  - Medical.
  - Demonstrations/disorder.
- Liaison with law enforcement agencies.
- Use of equipment
  - Weapons
    - Who authorized to carry.
    - What authorized to carry.
    - Prohibition on personal weapons.
    - Where to carry weapons.
    - Where and how stored; maintenance.
    - How to use weapons.
    - When to use them.
  - Radios
    - How to use.
    - Maintenance.
    - Codes.
  - Vehicles
    - Defensive driving.
    - Maintenance.
    - Records and reports.
    - Use of emergency equipment.
  - Portable public address systems.
  - Alarms
    - Where located/type.
    - Monitoring.
    - Records/reports.
    - Control systems.

- Response to alarms.
- Maintenance/service.
- Closed circuit television
  - Where located/type/capabilities/area covered.
  - How to use.
  - Monitoring.
  - Response requirement.
  - Records/reports.
  - Maintenance/service.
- Cameras
  - Use of department cameras.
  - Records/reports.
  - Accountability.
  - Maintenance.
- Fingerprint equipment.
- First aid equipment and supplies.
- Resuscitation equipment.
- Gas masks.
- Fire extinguishers
  - Types.
  - Use.
  - Locations.
  - Maintenance.
- Speed detection devices.

Obviously this is a large program. Too large to eat at a few sittings. Many subjects may not be appropriate to a particular school facility; additional subjects might apply at a particular school. The Security Director must constantly assess the interest, capabilities, and needs of his staff and blend these factors with his resources and operational time available to develop a continuing, interesting, and effective training program.

The Security Department should be located in a building assuring high visibility for the department. It should be easily identifiable and easily accessible to all individuals having business there, yet be safe and secure from vandalism or disorder. Because its facilities will be heavily used, and because high visibility is desired, it probably should be on the ground floor.

The department should be allocated adequate space for its administrative and operational requirements. Separate rooms or facilities should be considered for:

- Security Director's office (privacy for counseling).
- Branch administrators' offices (privacy for counseling).
- Conference room.
- Training room (combine with conference room).
- Interview and interrogation room.
- Communications room.
- Alarm monitoring area.
- CCTV monitoring area.
- Photography; identification photo area.
- Evidence storage.
- Weapons storage.
- Lost and found.
- Prisoner holding area.
- Expendable supplies.
- Supply storage; equipment storage.
- Files areas.
- Rest rooms, shower facilities, lunch areas.
- Medical treatment area (if not readily available elsewhere on campus).

Office equipment and fixtures, in addition to standard items (desks, chairs, typewriters, etc.), should include maps, photomaps, and blueprints of campus areas and buildings; photocopy equipment; some secured telephone lines to sensitive offices (Director, Intelligence, Operations).

Radio equipment should include a base station, mobile stations (police vehicles), and portable stations (walkie-talkies with patrolling Security Officers, at traffic posts, etc.). Radios should be equipped with battery chargers, and an adequate supply of spare batteries should be on hand at all times.

Radios should be capable of adequately transmitting and receiving from all parts of the campus. They also should have private frequencies. Sometimes campus security radios are on the same frequency as those of the school's maintenance crews or other service departments. This creates confusion, as when more than one station wishes to transmit at the same time and cannot; makes it difficult to maintain net discipline; and results in unauthorized disclosure of sensitive information. Security radios should be on a private frequency. Their use should be constantly monitored by the Net Control Station (usually the base station) which enforces communications discipline. The base station also should monitor radio traffic of local law enforcement and medical agencies for matters affecting campus security police.

Whether to arm or not arm Security Officers, the type arms, and their use are all the responsibility of the Security Director. Before making those decisions, he should consider their every ramification and consult with the school's head, its legal and financial representatives, and with local police officials.

Whether to arm or not to arm depends upon many things, not the least of which is the sensitivity of staff, faculty and students to the presence of armed officers on campus. Bring up the subject of armed police on campus and very quickly the shooting of students by an Ohio National Guard unit on the Kent State campus will come to mind. In most environments, it seems not necessary and not wise. What is worth the risk of a student's life to a policeman's bullet? On the other hand, urban campuses have constant problems with crime in its most violent forms, both from within and without the school. Has the administrator the right to deny a Security Officer, attempting to protect an individual from robbery, assault, or rape, a weapon to protect himself and to enforce that protection?

The best solution lies in a mix of armed and non-armed officers, as required by the particular campus situation. If an area or activity is particularly sensitive (storage and use of sensitive classified material on government projects, a high crime rate residential area, the emergency room of a university hospital) the patrolling Security Officers might be armed. Other campus Security Officers would not be armed. Security Officers armed with firearms, as opposed to batons, should never be used during student demonstrations.

Security Officers who will carry weapons must be carefully selected and well paid to insure their being the best possible individuals available. They must have undergone a thorough background investigation stressing their emotional stability. They must have satisfactorily completed state and local mandated training, fired the necessary qualification course, and met all other state and local licensing requirements. They must have fired the specific weapon they will personally carry.

Security Officers must never be permitted to carry their own personal weapons or any additional unauthorized weapons (leg-holstered extra revolver, knife, blackjack, chemical agent).

Weapons and ammunition should be the property of the school and be properly licensed with state and local authorities. They should be stored in the Security Department, issued at the start of a shift, and returned to storage when the shift ends. They never should be taken home.

If weapons are to be issued to Security Officers, the school must publish well-defined and strict policies limiting the drawing and use of those weapons to the protection of life. Under no other circumstances, should they ever be drawn. The policy should include provision for a full report by the officer and his supervisor of any time the weapon is drawn or fired.

Weapons, probably .38 caliber revolvers, must be carried in standard, school-owned holsters and use standard, school-provided ammunition. Under no circumstances should an individual be permitted to supply his own ammunition.

The existence and presence of weapons on campus require a continuing education program to convince the school community that the Security Department needs those weapons to better serve and protect.

Shotguns, sniper rifles, or similar weapons, if deemed necessary by the particular Security Department's requirements, should be kept in central storage for immediate issue only in emergencies, or be securely locked in patrol vehicles.

The use of batons (billy clubs, night sticks) also is a controversial issue. It is an excellent defensive weapon when an officer is faced by a dangerously disorderly or riotous crowd. A club, is a club, is a club, however, by whatever name it is called. So goes the charge of police brutality. As with other weapons, unless the area or activity is especially sensitive or subject to much criminal violence, batons usually are kept available in central storage for issue if necessary, but are not habitually carried by officers on patrol. If they are to be carried at all, the officers who are to carry them should be well trained in their use.

The use of chemical agents (tear gas, mace, etc.) has the same dangerous legal and moral implications as use of other weapons. They have little use on campuses but, if deemed necessary to have on hand, should be kept in locked storage.

Vehicles for use by the campus security force might include standard passenger vehicles, tow trucks (for traffic enforcement), and three-wheeled, weather protected scooters. High speed pursuits are not appropriate for a campus. It's better to lose an offender than to risk the lives of other drivers or pedestrians.

Police vehicles, except for those used by investigators, should be openly and clearly identified. They should be equipped with standard police accoutrements (spot lights, revolving police identification lights, sirens); and with emergency rescue equipment (medical supplies, fire extinguishers, flares, oxygen, pry bars, rope, cutting tools, etc). Weapons and protective helmets, if carried, should be locked in the vehicle when not in use.

Portable public address systems (bull horns) should be available in the department and in patrol vehicles.

Photographic equipment might include both 35 mm and polaroid cameras (investigations, traffic accidents, etc), polaroid photo-ID cameras (campus registration), hand-held television cameras, movie cameras, and portable spot lights.

Campus administrators debate whether their Security Officers should be "peace officers," fully authorized to make arrests and to conduct other police activities, or whether they should be simply a private guard force (contract or proprietary), similar to that in a business or manufacturing environment, with arrest powers only those of a private citizen.

The vast majority of campus police today operate under the same authorization as that of a private citizen.

The laws governing that authorization vary from state to state. In most cases, the private Security Officer can effect an arrest ("citizen's arrest") only when a felony (not a misdemeanor) has been committed in his presence or he has strong reason to believe that one has occurred or is about to occur and only his action can prevent it or apprehend the perpetrator. There must be no police officer available to make a valid arrest and, if the Security Officer makes a "citizen's arrest," he must deliver the perpetrator to a police officer as quickly as possible.

Problems? If the campus Security Officer does not see the "felony" occur, can he be reasonably sure that it has and that he has the perpetrator? Reasonably sure enough to win a legal suit for false arrest? If he is not police trained, can he recognize a felony and make a proper arrest (amount of force involved, search procedures, seizure of evidence, warning of accused's rights, detention until police arrive)? Once an apprehension is made, how long and in what manner can he "detain" the individual before police arrive? If he has not made an official "citizen's arrest," how long can he detain the individual before it automatically becomes an "arrest"?

In every instance the private Security Officer is personally vulnerable to criminal charges: assault, battery, negligent homicide (if he uses a weapon), false arrest. Both he and his school are open to civil actions: malicious persecution, defamation of character, false imprisonment, slander.

The defenses he offers may be reasonable enough if he were an average private citizen, but the campus Security Officer most likely will be expected to know and understand the laws of apprehension and arrest better than the average private citizen (and the school system will be expected to have taught him to know them and to properly supervise him in his duties).

In sum, the risks to the individual Security Officer (all of which affect the way he goes about his job), and to his school, in his having only the authority of a private citizen but being responsible for protecting the school and the school community, are substantial.

In many jurisdictions special police commissions may be conferred on campus security police. They must be authorized by the state and local community and this almost always involves other problems for the school administrator.

For Security Officers to be appointed peace officers, with power to arrest violators of public laws, the school probably will have to satisfy state and local authorities about such things as:

- There is a real need for arrest powers (size and type campus, incidence of crime, no local police assistance, etc.).
- The quality of the campus security force meets state and local standards (often stringent requirements as state or local law enforcement agencies won't want to be outdone).
- The campus security force is well trained.
- Supervision of the campus security force is excellent.
- School and Security Department policies are fair, clearly defined, legally defensible.

In most states these attributes may be hard to establish and maintain. They'll also be costly in terms of salaries, equipment, training, manpower requirements.

Other legal problems might include the following:

- Defining the jurisdiction of campus police (restricted to campus only or to some facilities off campus, matter of "hot" pursuit, etc.).
- Authority of campus police (*e.g.,* search and seizure).
- Degree of force authorized in making apprehensions and arrests.
- Handling of evidence seized on campus.
- Privacy of school records, to include campus police records.
- Working arrangements between local law enforcement agencies and campus police.
- Identification (uniforms/badges) of campus police.

The school will have difficulty obtaining and maintaining peace officer status for its security force. And, of course, the basic question is whether it really is necessary and worth the effort it involves. Probably not. In both cases. It depends, however, upon the school environment, its security problems, working arrangements among the school's administrators and local civic and police officials.

Again, a suitable compromise might be to have certain members of the campus security staff appointed peace officers, but not the entire staff. Appointed peace officers then would be available to assist others when necessary.

A similarly debatable issue relates to uniforms to be worn by campus police. Will they be police type uniforms or a less controversial, lower profile combination of blazers with distinctive pocket patches and coordinated slacks?

During and following the Vietnam War era there was a trend to avoid any appearance of a police unit on campus. Blazers were in vogue. More recently, however, because of far less concern for a police presence on campus and a greater concern about rising crimes in our schools, the mood has swung to police uniforms.

I believe a mix of both the police uniform and the blazer/slacks combination is best. It depends upon the campus environment and the specific campus activity involved. Officers engaged in traffic control or on patrol in high risk areas or activities involving potential disorder (rock concerts, athletic events), should be in uniform; those engaged in courtesy type duties (security receptionists) might well be uniformed in blazers.

Whatever "uniform" (blazers are a form of uniform) is authorized, it is critical that each Security Officer be neatly dressed in clean, good fitting, conservatively styled clothing. He'll be reassuring to those he protects, and he'll feel more confident about himself and his role in the school community.

No personal improvisations (different type shoes or ties, white sox, distinctive badges or patches) should be permitted.

Uniforms to be authorized to an individual vary in composition and number with the type duty, climate, etc., but should include both winter and summer attire in cooler climates, as follows:

- 3 long sleeve shirts; 3 short sleeve shirts.
- 3 winter weight trousers; 3 summer weight trousers.
- 1 all-season cap with rain cover.
- 2 winter weight blouses; 2 summer weight blouses.
- 1 raincoat.

- 1 stormcoat (depending on climate).
- 1 pair winter gloves (depending on climate); 2 pairs white or yellow dress gloves.

Whether shoes are provided by the individual or by the school, they should be uniform in appearance. Leather accoutrements and other equipment (whistle, handcuffs, flashlights, etc.) should be alike and should be worn in uniform manner.

Standards for proper wearing of the uniform should be clearly defined, published, and enforced at roll calls before each shift and by supervisors inspecting patrol stations.

Women have different body configurations (one of the many blessings of life) than their male counterparts and should be accommodated with uniforms that are similar in appearance to male uniforms but appropriate for female wear.

For the school to purchase, store, issue, repair, clean, and maintain accountability for uniforms is a very expensive proposition. Even so it may be necessary and/or more attractive, depending upon the local situation.

A number of commercial firms, however, do provide all these services on a lease basis. They should be carefully considered. They'll pick up all the administrative burden, to include repairs and cleaning, fitting and issuance to new personnel. It many instances they can do it better and cheaper than a school's attempting to do it on its own.

If cleaning and maintenance of uniforms are left to the individual security staff member, they should be handled by blanket payment agreement with a local commercial firm (or school facility) rather than by cash payment to the individual Security Officer. In most cases of direct payment, he'll not apply the allowance to its intended purpose and his uniforms soon will reflect his choice.

———

We have established a campus security department. We have given it a mission and assigned it a staff. We have given that security staff space and equipment for its various responsibilities, trained it, uniformed it. Now let us consider some of its basic operations.

CHAPTER 5

SECURITY OPERATIONS

THIS CHAPTER deals with a number of different operations performed by the campus Security Department. Much of the material suggested in this chapter will be developed in greater depth in discussions of specific areas of concern (fire prevention, security of residence halls, crowd control, etc.), but it is worthwhile to first consider the breadth of operations for which the department is responsible.

Police

The minimum number of Security Officers capable of covering a single post, 24 hours a day, every day, is five (allows for sickness, leaves, normal 40-hour week); a more reasonable figure, however, particularly for a small department where personnel flexibility is limited, is six.

If the school's staff is represented by a labor union, security staff members should be exempt from its membership. They should have all rights and benefits granted other employees but should not participate in union activities. Too often, union membership leads to conflicts of interest that can seriously undermine the department's ability to do its job.

The size of the campus police force should be determined by operational and budgetary requirements, not by a union contract requiring a specific number of employees or by any other factors.

Normally, Security Officers are assigned duties in three shifts: 8:00 A.M. - 4:00 P.M.; 4:00 P.M. - 12:00 P.M.; 12:00 P.M. - 8:00 A.M..

The hours causing most concern, from a security and public safety point of view, are 8:00 P.M. to 2:00 A.M.. After 2:00 A.M. the buildings are locked and/or secured by internal locks, roving patrols, CCTV, and alarms. It is possible to thin Security Officer coverage during this latter period to augment coverage during busier periods of the day.

44

Increased coverage also is possible by scheduling longer shifts or by using part-time security personnel.

If overtime or other costs for extra part-time guards or for overtime hours begins to creep up, it is time to consider adding a full-time guard to the staff. Neither extensive overtime nor extensive part-time help offers a good permanent solution. The individuals involved become tired of working extra hours or, if part-time employees, become tired of their limited use.

Campus Security Officers normally should not be assigned to two-man patrols. It is not necessary for them to patrol in pairs and the practice can lead to difficulties. If a lone officer needs help, he can get it quickly by his walkie-talkie radio or by an available phone. Two officers, patrolling in pairs, are more apt to talk to each other than to patrol and more apt to cover each other's sins of omission and commission (failing to properly check a residence hall, extra coffee breaks, napping on the job, etc.). Keep officers separated, require periodic communications checks, make supervisors constantly supervise their subordinates.

Supervisors should not be used for routine administrative tasks (answering telephones, handling the complaint desk, doing radio dispatches). Directing the performance of their subordinates is a full time job. Let a less qualified individual handle the department's routine administrative duties.

Patrolling should be a mix of mobile (vehicular) and walking posts. Vehicles quickly move a security force to an area where it is needed. They provide much broader area coverage and have a greater communications ability than that available to the officer on foot patrol. They restrict an officer's vision and hearing, however, and his ability to patrol particular areas (*e.g.*, residence halls) in proper depth and detail. Although police-marked vehicles will be seen by more people, in a limited area they do not inspire as much confidence as an officer on foot. Vehicle maintenance, particularly for a vehicle which is almost constantly in use, also is a factor to consider when weighing mobile *vs* foot patrols.

There is a place for vehicular (mobile) patrols on most campuses, but, other factors permitting (personnel, funding, campus environment), certain areas should be patroled on foot.

Assigned shifts, posts, and supervisors should be rotated periodically. Rotating posts and shifts enables the individual Security Officer to become acquainted with more than one post. It also allows him to become more acquainted with overall school activities and with more of the school community. Rotation also deters boredom and discourages Security Officers' forming mischievous associations. If different individuals patrol an activity, there is less chance of collusion between a guard and those who would misbehave.

Rotating supervisors allows Security Officers to broaden their experience by working under different individuals. It also provides for different supervisors' views of campus security.

If posts and shifts and supervisors are to be rotated, however, it should be done not oftener than every 4-6 months. More frequent changes are very difficult to adjust to and usually are counter productive.

All Security Officer assignments should be firmly established and posted, in advance, for all to see. Each Security Officer should be assigned to a specific post. It may be a fixed post (as a security receptionist in the university hospital or at a campus traffic station) or a mobile post (as in vehicular patrol of residential areas).

Each post should be clearly and thoroughly described in its Special Orders. Special Orders describe everything unique to a post, as opposed to General Orders which cover all posts and general security matters in more sweeping form.

Special Orders, prepared for each post, should describe that post in sufficient detail so that a Security Officer new to it would know:

- The specific location and boundaries of the post.
- Type post (fixed or mobile).
- Hours to be manned; number of guards.
- School activities at that post.
- Presence and control of alarms, CCTV, other electronic security measures.
- HVAC, electrical, water controls for emergencies.
- Uniform to be worn; any special equipment.
- Fire fighting equipment (type, location).
- Particular requirements for the guard at that post.
- Reports required (communications checks; vehicle, package or visitor logs; traffic surveys, etc.).

There should be a slight overlap on each shift to assure that outgoing Security Officers properly brief their replacements and that incoming officers are inspected by their supervisors. Fifteen minutes' overlap should be ample. Hopefully, this overlap can be accomplished without additional compensation but, if 15 minutes are required to insure a neat, clean, alert, knowledgeable Security Officer on post, and that means 15 minutes' extra pay, it's worth it. Supervisors, of course, must insure that the extra time is used for that purpose and not for idle gossip or an extra coffee break.

Once equipment has been transferred and information exchanged, remaining extra time might be used for the outgoing Security Officer to complete and submit his shift report and for the incoming Security Officer to begin his own report. Reports, to be discussed later, should be entered chronologically as soon as possible after patrol events occur and, in every instance, must be completed and submitted by the outgoing Security Officer before he goes off duty.

It is better to relieve guards on post than to bring all to a central station. The latter approach leaves posts uncovered and wastes time. And exchange of information about a post is more effectively done where situations described actually exist.

In all cases, incoming Security Officers should sign in and be given a quick briefing and inspection by their supervisors (appearance, equipment, fitness for duty) before going to relieve the outgoing Security Officers on their posts.

The Security Officer's personal equipment should not include a massive collection of keys. Chances are that many of them no longer are valid anyway (lock changes, locks no longer used, not appropriate to the post he now is assigned). The more keys he carries, the more he is likely to lose or to carry home rather than pass on to his replacement. Particularly master keys.

Unless they really are needed on post, keys should be secured in the central headquarters for issue only when actually needed and only to the appropriate Security Officer.

All Security Officers should carry a small, personal field notebook. It should be loose leaf to permit easy insertion and removal of pertinent material and inspection by supervisors. It should be small enough to fit in a uniform without causing an unsightly bulge.

The field note book should contain only information he may need on the post to which he is assigned, such as:

- Telephone numbers.
- Fire alarm locations, codes.
- Radio codes.
- Incident classifications (file reference purposes).
- Special Orders for that post.
- Information on specific incidents he is likely to encounter (special school events, traffic problems, etc.).

In sum, the field notebook should guide his performance on post and contain material on which he can record observations and facts concerning conditions or incidents he finds on that post.

Too often, personal notebooks contain general information (general purpose of security, benefits program, etc.) he should have available to him, but not on post. It is not needed there. It should be recorded in a much larger loose leaf Security Operations Manual. This manual, issued to each Security Officer, would not be carried on post. It would contain such general information as:

- Mission of the Security Department.
- Table of organization.
- Personnel roster; addresses; phone numbers.
- General Orders; Special Orders for all posts.
- Training information.
- Sample security records and reports.
- Personnel policies.
- Uniforms (authorization, proper wear, maintenance, accountability).
- General patrol procedures.
- Equipment (authorization, accountability, maintenance, use).
- Weapons (authorization, accountability, maintenance, use).
- Fire procedures.
- Medical emergencies.
- Safety procedures.
- Disaster control.

Field supervisors should periodically check both the Security Operations Manual and the Security Officer's personal field notebook to insure that they are being properly maintained (currency, etc.), that they hold only current information, that they are being properly used.

At least once during each shift supervisors should inspect every Security Officer on post. The supervisor should be concerned with the Security Officer's uniform and equipment, his personal appearance and behavior, and the appearance and condition of his post. Supervisors should record these inspections in their own shift reports and use the time with individual Security Officers to teach and motivate. Personal leadership at all supervisory levels is key to the success of any security force.

Periodically, but at least semi-annually, supervisors should counsel the Security Officers they supervise on their overall job performances. Annually, this should be done in writing with written endorsement by the next higher supervisor.

Traffic and Parking

Certainly one of the most challenging responsibilities facing the campus Security Department is control of campus traffic and parking.

Faculty and students alike, it seems by their very natures, generally are rushed—or think they are—and won't appreciate delay getting to class at the last possible moment, leaving it at the earliest. On a busy campus, vehicle traffic may be heavy and rushed. Add to that the safety hazards posed by motorcycle and bicycle traffic and pedestrians scurrying to evade them all. It's more than enough to cause the Security Director to linger over a second cup of coffee.

Major concerns of the security staff in controlling campus traffic include:

- Planning for traffic control (present and foreseen problems); conducting continuing traffic studies to improve the program.
- Proper use of electric traffic signals.
- Proper use of Traffic Officers.
- Keeping traffic moving in orderly fashion, as quickly and as smoothly as safely possible.
- Controlling and assisting traffic flow in congested areas.
- Keeping fire lanes open.
- Assisting emergency vehicles.
- Assisting in accidents: evacuating casualties, clearing roadways, keeping traffic moving.
- Investigating and reporting traffic accidents.
- Assuring the availability of emergency support when required (towing, fire, medical).
- Insuring safe and orderly parking.
- Assuring firm, fair, impartial, quick, professional enforcement of traffic regulations.
- Supervising vehicle towing and impoundment.
- Conducting special motor vehicle and motorcycle or bicycle safety classes.
- Providing traffic control training for the security staff.
- Providing VIP escorts.

These are continuing duties but the circumstances surrounding them constantly change. Variations in traffic, peak periods, weather, campus activities, and other factors may sharply affect the best way to meet all these responsibilities.

The Security Director must constantly evaluate traffic conditions and be prepared to shift his resources (manpower and equipment) to meet them. Flexibility, adaptability, imagination, and the ability to stay calm when others are getting excited are all worthwhile attributes in a Security Director.

Basic procedures for traffic control, however, should be written and taught to all security staff members who will implement them.

A subsequent chapter discusses the detailed traffic and parking studies necessary to develop these SOPs; elements of traffic and parking SOPs; and their operational implementation by the campus traffic staff.

Investigations

Add to the growing list of campus security responsibilities the conduct of investigations.

"Investigations" includes responsibility for the assignment of trained investigators to thoroughly research certain matters; supervision of the investigative process; the evaluation of completed investigative reports for accuracy, clarity, thoroughness, completeness, and legality; recommendations suggested by the investigation; and the protection of information gathered during the investigation.

Campus investigators should be among the best qualified and most experienced members of the Security Department. They should be noted for their maturity, ingenuity, persistence, perceptiveness, fairness, discretion, and ability to work well without close supervision.

Matters subject to investigation by the school's security staff might include the following:

- Crimes and allegations of crimes committed on campus.
- Vandalism or other mischevious damage or destruction of campus property.
- Complaints against the Security Department.
- Theft, misappropriation, or misuse of school property.
- Violation of school rules and regulations.
- Personnel background investigations.
- Suitability of current security and safety procedures.
- Accidents.
- Unsafe conditions.
- Potential or actual liability claims.
- Allegations of student involvement in drug abuse, gambling, improprieties involving the school's athletic program.

Investigations by the campus security staff are discussed in detail in the next chapter.

Safety

An additional responsibility of the Security Department is planning and executing the school's safety and emergency services programs.

Safety and emergency services fall naturally to the security staff. They are trained in these matters; they are present on campus in most places, at most times, every day; their department is the center of the school's communications capability.

Just as for other security matters, security staff members would conduct safety surveys and inspections and recommend measures to improve existing conditions. They will investigate accidents, fires, and other manmade disasters. They may direct or monitor medical services, fire prevention, and fire fighting programs. Considerably more on these subjects later.

Miscellaneous Security Services

Additional miscellaneous services which might be assigned to the campus security staff include:

- Providing Security Officer escorts for staff and faculty members or female students during high crime incidence hours.
- Providing cash escorts or depositing cash receipts for after after-hours bank deposits.
- Clearing cash registers; taking register tapes to the department designated to process them.
- Providing emergency mail or messenger services.
- Checking parcels and packages.
- Receiving parcels, packages, messages, incoming shipments.
- Distributing reports.
- Providing escorts for visitors.
- Performing courtesies to national flags on campus.
- Handling photo-ID cards or badges (preparation and issuance, recovery when individual no longer required to have them or no longer authorized).
- Providing key services for opening secured areas for authorized individuals.

All of these services may be assigned to the Security Department. All can be valid; all can provide a real service to the school. The danger, however, is that they may divert officers from their primary mission of protecting the campus and the school community. If these miscellaneous security services can be performed without jeopardizing the officers' primary duties, they may be acceptable.

Once these courtesies are extended, however, chances are it will be difficult to recall them and the individuals benefitting from them probably will want them extended. Then security can become of secondary concern and Security Officers little more than campus messengers.

The Security Director must always keep his fingers squarely on the pulse of the department. The moment he sees the Security Department's primary mission in jeopardy, he must take action to prevent that happening.

Operations of the Security Department should embrace every physical facility or activity on the campus, cutting across all departments.

Unlike some departments, such as admissions or athletics, security operations produce no tangible "revenues" in new students, new scholarships, new buildings or equipment. Normally, the more low-keyed the operation, the better. This often creates difficulties for the Security Director fighting for his share of the school's resources so his department can do its job. The school community simply is not aware of the many services the security staff provides or how critical they are to the school's health.

Hopefully, this chapter will cause its readers to be more aware of those matters than they were before.

CHAPTER 6

CAMPUS INVESTIGATIONS

IF THE SCHOOL has no security staff, a student's threatening a teacher, the theft of a tape recorder from the school library, or vandals' flooding the gym locker room may be reported to a department head. Then it may be discussed at the next staff meeting. But it's not likely that any action will be taken about it.

If the school has a Security Department but no investigative staff, the incidents may be recorded in the logs of the Security Operations Center. Probably no further action will be taken to apprehend the perpetrators, however, because there is no one to take it.

If the school reports the incidents to local police, in a few days a local detective may visit the campus, copy the Security Operations Center's log describing the incidents, perhaps even look at the incident sites. Unless the crime is a serious one, however, the investigation probably will end there. The local police department has enough problems of its own without taking on those of a less serious nature on campus.

The campus Security Department needs an investigative staff. It may have to be an additional duty for one individual, but someone should be available to conduct campus investigations.

Depending upon the size of the security staff, 2-3 individuals should be designated as investigators. The Security Director should negotiate an informal arrangement with local police so that those individuals will investigate less serious offenses on campus, coordinating their work with local police when appropriate. More serious crimes must involve local law enforcement agencies.

Individuals designated as campus investigators should be mature, trained in investigative techniques, experienced in conducting some form of investigations, and have good "people skills." They also must be discrete. Their work will involve many individuals associated with the school. They will become privy to sensitive information about the school, about the school community. They must know to keep that information to themselves.

In selecting individuals for this sensitive position, the Security Director (or school administrator if he has no formal director or department), might consider that solely in investigating a reported incident that investigator might become involved in:

- Assisting injured persons.
- Locating, identifying, and interviewing participants and witnesses to the incident.
- Collecting and properly protecting evidence.
- Collecting and recording all facts (written, verbal, photo) pertinent to the incident.
- Completing accurate, thorough, complete, and comprehensible reports of all facts related to the incident or its investigation.
- Identifying, apprehending, and possibly arresting perpetrators.
- Coordinating with local law enforcement agencies.
- Testifying in criminal or civil courts regarding the incident or its investigation.

Investigators should wear civilian clothing, rather than campus security uniforms, and be equipped with excellent sets of badges and credentials.

Incident Investigations

On every school security staff, at least one person should be designated to handle complaints and to conduct limited investigations of pertinent campus matters. Perhaps he will never determine who is stealing private property from the mens' dorm, but the very fact that following a complaint he appears, asks knowledgeable questions, possibly makes changes that augment the security of the affected area, and later gets back to the victim to let him know that he has done *something*, will all have a beneficial effect.

Should that individual(s) not be specifically trained in investigative matters, the following suggestions are offered.

In handling complaints, remember that the complaint probably is important to the person making it. Listen. Listen quietly and attentively. Don't interrupt unless absolutely necessary. Save your questions until he has said what he has to say. The first time around, keep note taking to a minimum. When you review the complaint with him, make detailed notes. Later, you will seek individuals who can verify details of the complaint. Don't argue. Be understanding but reserve judgment. Don't offer an opinion that can come back to haunt you. Assure the complainant that you will be discrete in using the information he is providing and that you will do something about his complaint. Later, give him some report on the results of your inquiry.

If the incident involves a crime, on arriving at the scene, the following actions might be appropriate:

- Arrange for necessary first aid; evacuate victims requiring medical help.
- Assure the safety of the area (fire, etc.).
- Secure the crime scene area from anyone not required for the investigation. Allow nothing to be touched.
- If appropriate, notify local police.
- Study the area carefully for physical clues, making detailed notes and protecting any physical evidence recovered.
- Conduct witness interviews.
- Identify and interview additional witnesses who may have left the scene.
- When ready, release security of the incident area.
- Determine and follow other investigative leads.

In interviewing a witness to an incident, keep the following ideas in mind:

- Prepare for the interview. Do your homework. Study the complaint and any other developed information. Do that before you interview anyone, not during the interview.
- Be certain that your own appearance is neat, clean, attractive.
- If possible, choose the time and place for the interview. Face-to-face interviews always are preferable to telephonic interviews.
- Determine whether someone else should be with you for the interview. For example, if the witness is apt to be hostile or if it is a woman and the subject may be offensive to her, you may wish to have another officer along to help and to witness the interview. Usually, however, it works out best to have as few people involved in the interview as possible. The interviewee is more apt to confide in a single interviewer than to several participants.
- Allow adequate time for the interview. Don't rush it. It takes time to establish rapport (your first objective) and time to develop information. Then you must go over it again to be sure that you have it correct.

- Be sympathetic, understanding, calm, patient, friendly, but firm. Remember that, at this point, you are not dealing with a wrongdoer.
- First establish rapport, common interests. Once the witness tells you anything, you can begin to steer him toward your topics.
- Control the interview. Hold to the subject you want to talk about. Listen. Don't be taking notes, or thinking of your next question, and miss something important. Watch his gestures, mannerisms, sensitivity to particular topics, etc.
- Phrase your questions to get a narrative type answer, not a simple "yes," or "no."
- Watch for discrepancies (they may not be a lie, simply a discrepancy in his understanding or in the story as you've heard it from another witness); return to them from another angle; try to clarify them.
- Be thorough. Get all the details. Watch for vague expressions like "Oh, you know what I mean." You *don't* know what he means. Make him tell you; then you have it straight.
- Look for witnesses to collaborate his story.

If the witness is hostile, but his story worth getting, cultivate him. If he is lying, attempt to get the truth by pointing out discrepancies in his story and going over it again and again from different angles. If he persists in lying, try to question him under oath and get a sworn statement. Later you can go back to him with it.

Some witnesses attempt to question the questioner. Don't tell them all you know or say anything that may come back to haunt you. Besides, you're only wasting time. If appropriate, give him a little information, implying that you know much more. Use that as a lever to get the information you want.

Interviews with individuals suspected of having committed a crime have several objectives:

- Establish the guilt or innocence of the individual.
- Discover motives, methods used, extent of the offense.
- Identify any others who may be involved.
- Obtain legal evidence or leads to legal evidence.
- Obtain a written confession, signed and witnessed.
- Seek recovery, replacement, or repair (where appropriate) of property involved in the crime.

Suspect interviews are the most difficult to handle. They go better with experience, but no two are quite the same. The following suggestions may help:

- Give the individual the benefit of the doubt until he takes the doubt away.
- Unless you are absolutely sure he has committed the offense in question, open with an indirect observation, *e.g.,* "We have some irregularities to be explained," or "Some violations have occurred," or "We need your help to set a problem straight."
- Don't paint an all-black picture; offer him a way out of his dilemma.
- As you question him, keep the suspect's viewpoint in perspective, leaving him a face-saving escape if possible (bad company, we all have debts, etc.).
- Work toward a written statement of his involvement in the matter being investigated.
- Avoid inflammatory words: "Jail," "Thief," "Police," etc.

Obviously these suggestions will not make an untrained individual an expert investigator. They are valid tips, however, and might well be considered if your own experience in that area is lacking. A number of good textbooks on investigative techniques also are available. The best guide, however, is to work deliberately, carefully, intelligently. Treat the people you interview with courtesy. Listen carefully. Resolve any questions that come to your mind. As soon as possible after the interview, use your detailed notes to prepare a formal investigative report. Review that report before you do subsequent interviews concerning the same incident. Retain your notes for future reference should you ever be required to testify about the investigation.

Do all this and you may be surprised that your investigations can work out rather well. And one apprehended thief or vandal, one instance of recovered goods, can make all the work worthwhile.

Background Investigations

Background investigations verify the accuracy of data provided by a job applicant and may determine his suitability for employment, promotion, or assignment. They are better conducted by the school's investigative staff than by a contracted investigative agency.

The campus Security Department can do it better, faster, and more economically. And the Security Director can know exactly who has been contacted and what that individual said.

Most contracted investigators, for example, summarize their investigative reports with a vague: "Seven sources were contacted. No adverse information was developed." What sources? Who were they? How well

do they really know the subject of the investigation? Were they his rela-
tives, friends, business associates? Were they biased? What questions
were asked? Was the "interview" by mail, by phone, by face-to-face interviews? It's easier to avoid giving detrimental information if the questioner is calling by phone rather than face-to-face with you.

If the background investigation is conducted by his own staff, the Security Director can better evaluate sources and the information they
provide.

If possible, all school employees should be subjected to some type
background investigation. Its depth and overall scope would be determined by the sensitivity of the position involved, but data provided by
all staff and faculty members should be verified and, where appropriate,
those individuals' suitability for prospective appointments should be assessed by verification of their personal attributes and work habits.

Background investigations are among the most sensitive undertaken
by campus investigators. They are going to find discrepancies between
claims and truth, deviations in personal behavior. These matters can
lead to hard decisions by administrators, particularly as they involve
staff members. Making those hard decisions during the employment
processing, however, can save the school money, time, and embarrassment in the long run.

Intelligence Collection

A type investigative role played by all members of the Security Department is the collection of information about campus activities. This
information, properly reported and properly evaluated, creates intelligence which can forewarn administrators of such potential problems as
the presence of drugs or gambling on campus; theft, damage, or misappropriation of school property; or violations of school regulations and
procedures.

The best gatherer of this type information is the individual Security
Officer. As he works with the school community, his honest conversation, perceptive questioning, even more perceptive listening are enough
for him to earn the confidence of staff, faculty, and students. Confidence
which results in shared information which may not be available through
routine departmental or school administrative staff meetings.

There is nothing legally, ethically, or morally wrong with this form of
information gathering and reporting. It can save the school much trouble, particularly when problems of a serious nature are developing and

this form of "early warning radar" leads security investigators to research the causes and extent of the problem so the school administration then can take measures to prevent or lessen a worsening situation.

Surveillances

Campus investigations may include surveillances. They may be overt (as at a traffic post) or covert (as in an undercover investigation on drug abuse). They may include honesty shopping at the school gift shop or spot inventories to verify school purchases.

Covert surveillance is a limited weapon available to the Security Director. If successful it can save the school a lot of money (as in identifying individuals involved in theft of school property). It also can lead to significant intelligence on school disorders or evidence of criminal activity.

A form of covert surveillance is the undercover investigation. It can be an effective tool in investigating illegal activities (*e.g.,* identifying drug pushers) and concurrently can provide legitimate information on such matters as job performance, staff morale, attitudes toward supervisors.

It is a very expensive form of investigation, however, and very sensitive. Administrators approving an undercover investigation must realize that the undercover operative may never be able to gather the evidence they want and may himself become involved in the illegal activity he is investigating to win the confidence of those involved in it. And, it goes without saying, that Americans dislike any form of covert inquiry. The school administrator must be ready to face that issue if the undercover operation is exposed on his campus.

An undercover investigation may require five to six months' work before it begins to bear fruit. When it ends, the undercover operative usually must be extracted from the operation, and perhaps from the school itself. This is to protect him and to safeguard his future usefulness. It also is done to prevent disclosure that an undercover operation ever was conducted on the campus. Occasionally it is possible to use an individual from a particular school in an undercover capacity at that same school. More often, however, the Security Director will employ an individual contracted from a contract security agency or "borrowed" from the investigative staff of another school. In any event, his identity and activities must be protected. If they are discovered, he should be removed from the campus.

Undercover operatives should submit a verbal report to their supervisor as soon as possible after significant information is developed. Written reports should be submitted at least weekly. The supervisor should review these reports with the operative and use them, and the debriefing sessions, to further target the undercover investigation.

Reports of undercover investigations are particularly sensitive. They should be discussed only with individuals *authorized* to know that information and who really need to know it to do their jobs.

There is a human weakness to share a juicy tidbit of information or a bit of gossip. Those urges must be resisted. Unauthorized disclosure of an undercover investigation can cause great embarrassment to the school. More important, it can get investigators or others involved in the inquiry hurt.

Security Surveys

A final example of a form of security investigators' efforts is the conduct of security surveys and inspections to determine security and safety hazards and to recommend measures to eliminate them. This is covered in depth later but, indicative of the experience, maturity, good judgment required of the investigator conducting security surveys, he would be required to:

- Research the purpose and organization of every school department being evaluated and existing departmental procedures.
- Develop a list of security or safety threats to that department.
- Assess existing safety and security measures.
- Recommend additional or revised safety and security procedures.
- Provide oral briefings on the results of his surveys for department heads concerned.
- Prepare written reports of his surveys.
- Conduct follow-up inspections to determine the presence and effectiveness of amended or new procedures.

In accomplishing these things the security professional must touch the face of every school department, every school activity, every key individual involved on the staff and faculty. He will be privy to sensitive and valuable information and must use great discretion, maturity, and sound judgment about using it.

One sound recommendation, as for fire prevention in a residence hall, can save lives, prevent damage or loss or school property, or avoid a law suit that could equal the investigator's salary for his entire career. His is a very important job and should be treated that way.

Investigations are important to a school's security program. Even if the school has no formally constituted security department, one or more individuals should be trained in investigative techniques and assigned responsibility (and necessary authority) for investigating appropriate matters brought to the school's administrator's attention. Their capable inquiry now may save that school much money and much embarassment just a short way down the road.

CHAPTER 7

TRAFFIC CONTROL AND PARKING

ONE OF THE greatest challenges a school's security force must face is the proper control of campus traffic and parking.

Campus roads and parking lots are filled with vehicles. Drivers rush to and from campus activities; pedestrians scurry to cross roads safely. The Security Director is charged with insuring a rapid, smooth, safe flow of vehicular and pedestrian traffic and with assuring secure parking near campus facilities. Doing it involves the whole school community.

Because of that involvement, a committee including representatives from staff, faculty, and students should be involved in overall campus traffic and parking planning and enforcement.

Most of the work in developing this plan, certainly in enforcing it, however, will be done by security staff members. They'll begin by evaluating existing traffic conditions and control measures.

They might start with traffic signals. Signals that regulate and guide traffic. To do that properly, signals must be located at intersections where they can do the most good; they must be a type appropriate to that intersection; they must operate satisfactorily (as a unit and in concert with adjacent signals); and they must be clearly visible.

Each campus electric traffic signal should be studied to determine that:

- It is appropriately located.
- It is the type signal appropriate for that location.
- It is clearly visible to approaching vehicles.
- It is in good repair.
- It is not unduly vulnerable to damage through accidents, vandalism, or weather.
- Its directions (signals) are clear and accurate.
- Each phase of the signal is properly timed and is coordinated with adjacent signals.

Traffic surveyors then might assess the proper use of road traffic signs:

- Are road signs placed everywhere they are needed?
- Are road signs appropriate for their location?
- Is every road sign necessary/helpful? Are some of limited necessity/help? Are any distracting to drivers?
- Are road signs in good repair (readable, free of significant damage, reflectorized)?
- Are road signs clearly visible (free of grass, trees, shrubbery, etc.)?

Speed studies might come next. They should be conducted at the same times during three periods (*e.g.*, 9-12, 3-6, 8-10) every day during the assessing period. They should be for a uniform period (*e.g.*, 1 hour or 50 vehicles each way).

Through these speed studies the planner can determine such things as:

- The validity of campus speed limits.
- Proper speed limits.
- Proper speed limits for new routes.
- The effectiveness of electric and human traffic control measures.
- If and where additional or different traffic control measures are needed.
- The relationship of speed to accidents on campus.
- The effectiveness of the overall speed control program.

A variation of speed studies considers delays in traffic and may reveal such things as:

- Overall speeds typical on campus.
- Typical moving speeds from one point to another.
- The necessity for road construction or improvements, as in insufficient road capacities.
- The effect of road conditions on traffic.
- The effectiveness of traffic control measures.
- Adequacy of speed limits.
- Variations in traffic during peak and non-peak hours.

Still another form of speed-delay studies is concerned with origins and destinations. Normally it is conducted by interview at campus traffic check points but may be in the form of written check lists sent to selected members of the school community. The latter can be done by post card type forms or a questionnaire sent out with registration forms. The problem with the "post card" or volunteer approach is that most individuals, unless they are personally interviewed, won't take the time to answer a check list or at least to answer it completely and properly.

Vehicle operators chosen for speed-delay studies might be interviewed at the campus gate or another marked point. Concern here is with travel to and from the campus from an off-site location or solely with traffic between points on the campus.

Speed-delay questionnaires might ask:

- Where did the trip originate?
- Where is your destination?
- What gate did you use?
- Which streets (routes) did you use?
- Where do you normally park on campus?
- How many people are in your car? Is this normal?
- If a car pool, what other stops are made?
- What is the purpose of the trip?
- How long does the trip normally take?
- What problems are you apt to encounter? Where? When?
- Do all traffic signals operate satisfactorily?
- Are Traffic Officers effective?
- Have you any suggestions to correct or improve traffic conditions?

Use of traffic check points can provide additional traffic control data. Not too popular in the school community when they result in traffic citations, they are still a valid tool in campus traffic control. These are Traffic Officer-manned portable stations periodically operated (random selection of areas, locations, primary focus of security interest) to determine, among other things:

- Proper use of routes.
- Inspected vehicles are properly registered and safe to operate.
- Inspected vehicles do not contain contraband or other illegal property being transported on campus.
- Inspected vehicles, their operators and passengers, have a valid reason to be on campus.
- Vehicle occupants are not under the influence of drugs or alcohol.
- Operators are properly licensed and are complying with school regulations.

The following suggestions apply to traffic check points:

- They should be hidden from distant view, to prevent by-passing, but not appear as a sudden surprise.
- The check point should be off the main roadway.

- Lanes approaching the check point should be clearly outlined with traffic cones, tape, and signs.
- Barriers should be provided to stop vehicles at the appropriate point.
- Traffic Officers should alert drivers of vehicles to be checked early enough to allow them to safely enter the check point area.
- A patrol vehicle should be available to pursue noncompliers.
- The selection of vehicles must be consistent (*e.g.*, all, every odd vehicle, every 5th vehicle, etc.).

Several additional campus traffic studies might be worthwhile. The first is a study of vehicle registration files to accomplish the following:

- Update origin-destination studies.
- Assist in planning parking facilities.
- Identify vehicles registered on campus by year, make, model, color, license number, ownership, etc.
- Identify stolen, abandoned, or recovered vehicles.
- Identify individuals of particular interest to the security staff.

The second is a study of accident records, useful for the following:

- Identify high accident rate locations.
- Identify common denominators in accidents, leading to necessary corrective action.
- Evaluate roadway design and condition, leading to necessary corrective action.
- Evaluate existing traffic control and parking programs.
- Suggest other remedial actions, such as selective enforcement programs.

Through all these traffic control studies, planners hope to improve traffic flow, as by:

- Redirecting some traffic to other routes.
- Correcting phasing of electric traffic signals.
- Encouraging staggering of work hours (class changes, etc.) with a 15 minute separation between groups.
- Encouraging car pools.
- Improving and encouraging use of public transportation.
- Restricting the use of certain vehicles (or roadways or times of use).

In planning traffic control measures, they should keep in mind certain basic principles:

- One or more major routes should connect main highways to the school's major functional areas.

- When possible, campus traffic should be concentrated on a limited number of well controlled and clearly marked routes.
- Campus traffic should be directed to but around, rather than through, heavily congested major functional areas.
- For special events:
  - There should be direct routes to and from the site of the event.
  - These direct routes should begin and end at parking areas.
  - Special routes or lanes should be marked for emergency vehicles.

Traffic Officers

The appearance and conduct of Traffic Officers are critical to their value to the traffic control program. Former Military Police veterans invariably are well trained for this task. They also usually make excellent supervisors and teachers for less experienced officers.

The following suggestions, by no means all inclusive, apply to Traffic Officers:

- Key to success in controlling traffic: *Look* carefully at the developing situation; *Execute* the appropriate signal properly and concisely; *Complete* each signal distinctly.
- Always execute clear, distinct signals; don't get sloppy in your thinking or in your signals.
- If the driver does not seem to understand, repeat the signal.
- Carry yourself erectly but relaxed at all times; this reduces fatigue and presents a good appearance that commands respect.
- Be certain that your uniform is clean, neat, complete, comfortable, correctly worn. Your badge should be prominently and properly displayed. Wear gloves to accentuate hand signals.
- Body language: important. Traffic Officer stands with his side parallel to the direction of traffic flow, his front and back toward stopped traffic.
- Whistle: use it; it establishes command presence. Properly used, it helps a "cop" become a Police Officer. It can be heard above other sounds.
- To start traffic: two short, stacatto blasts, accompanied by hand signals.
- To stop traffic: long, continuous blast, accentuated by hand signals.
- To get attention: series of short, stacatto blasts, accompanied by hand signals.

Enforcement

Key to any traffic control plan is enforcement. A poorer plan is made much better if everyone complies with it. There will be order. Perhaps not the best order, but order. That's a help.

Security Officers, particularly those involved with traffic control, must be taught to detect and to identify criminal behavior and to separate it from garden variety relatively harmless misbehavior. They also must be taught, and reminded from time to time, that professional demeanor and courtesy will buy a lot more on a campus than will their badge or uniform alone.

They must be fair with all violators: take the same action for the same violations under similar conditions. The average school staff member or student or visitor is not a criminal. He'll violate regulations, intentionally or unintentionally, but most individuals are not criminals with total disregard for public order. Police officers must maintain a certain reserve, a distance from those they may have to force to do something the individuals really do not want to do (obey rules). But they still can be friendly, polite, outgoing.

When a campus Security Officer or Traffic Officer must give a driver a warning or citation, he should start with a pleasant, courteous greeting. This normally will disarm an angry, frightened, or embarrassed driver and defuse a potentially explosive situation. Particularly if the driver is not alone in the vehicle. To avoid unnecessary embarrassment to the driver, the officer should remove him from his family then correct him in a firm but controlled voice and manner. Anything less would be unprofessional and more likely to cause a confrontation.

The traffic citation for use on campus might be a three-ply document issued in books of ten (10). Tickets would be numbered and require strict accountability. Traffic Officers should never collect, accept, or otherwise keep track of fines paid or unpaid. They should take no action concerning nonpayment; that is a matter for others in the school administration to handle.

When a citation is issued, distribution might be as follows:

- 1st copy (white): Return to Security Department for transmittal to the school office responsible for processing citations.
- 2nd copy (yellow): Security Department's file.
- 3rd copy (envelope): Issue to violator (or place on windshield).

Concerning the form itself, all violations require the following data:

- Traffic Officer's identification number.
- Date of citation.
- Description of violation, including date, time, place.
- Type violation and location code numbers (quick file reference).
- Vehicle description (make, color, type).
- License plate number and state.
- Driver's name and status (faculty, student, visitor), if available.
- Driver's address (if available).
- Driver's license number (if available).

If the offense warrants appearance before a school authority, that information, with complete time, date, location of appearance, should appear on the form. There also should be a provision for requesting voluntary appearance before that authority.

Enforcement of traffic and parking rules should be equitable for everyone. If traffic citations are to be issued, all offenders (staff, faculty, students) should be equally targeted and pay equal fines for similar offenses.

Any appeals system for traffic citations also must be fair and equitable. Its appeals board should include a representative of each element of the school community. The appeals system should be kept as simple in concept and operation as possible, never allowing an administrative delay (as in infrequent appeals board meetings) that discourages realistic punishment.

Assessed fines must be severe enough to cause some hardship to the recipient and financially worth the cost of administering them. They also must be severe enough (*e.g.,* $3, $5, $10) for successive repeat offenses to cause an individual to think twice before he repeats an infraction.

All this should be spelled out in the school's traffic and parking regulations. Those regulations also should include a set limit on the number of violations permitted before the offender is required to undergo a driver's training program or have his right to drive on campus permanently rescinded.

Accidents

Campus Traffic Officers must render aid after vehicle accidents, complete on-site traffic investigations and reports, apprehend (where appropriate) and issue appropriate citations to individuals involved in the accidents.

Security vehicles should be equipped with the following equipment for accident emergencies:

- Blankets.
- Flares (red signal).
- Portable reflectors or flashing lights.
- Engineer tape.
- First Aid kits.

- Flashlights.
- Warning flags and stanchions.
- Reflectorized vests.
- Fire extinguishers.

All security vehicles should be equipped with a spotlight, siren, flashing lights, and a public address system. Security vehicles used for rescue purposes also should be equipped with pry bars, heavy duty scissor type jacks, 25' tow chain or cable, shovels, axes, wrenches.

Traffic Officers on the scene should first secure necessary medical and fire (if appropriate) aid and evacuate all injured personnel. They must insure that the road is properly marked with flares or other signals to warn impending traffic of the accident area. They must see to the earliest possible removal of debris from the roadway and the smoothest possible resumption of traffic flow around the accident.

A broad variety of types of accident report forms are available for the Security Director's consideration from local police, other schools, insurance agencies. Whatever form he develops for his school's use, it should contain provisions for an accident sketch. The sketch should clearly identify:

- Roadways, shoulders, other critical areas of the accident scene.
- Appropriate signals or other control devices.
- Base points (permanent objects such as telephone poles, signs, buildings, etc., useful as references in describing the accident).
- Start and final positions of vehicles involved.
- Information on vehicle marks and debris.

When the accident sketch is completed, it should be carefully rechecked by another officer for completeness, clarity, and accuracy.

It also would be helpful to obtain reports from the drivers involved in the accident. A form similar to that shown in Figure 2 would be appropriate.

Besides data concerning individuals immediately involved in the accident, names and addresses of other witnesses should be reported.

DRIVER'S ACCIDENT REPORT

Notice to Drivers: This is your report of what happened in the accident. Answer each question completely.

1. I was driving _____ on _____.
   (direction)   (street, town)

2. I first saw the other car when it was _____ away from me.
                                          (feet or car lengths)

3. I was going _____ at this time.
   (miles per hour)

4. The other car was going _____ at this time.
   (miles per hour)

5. My car went _____ after the impact.
   (feet or car lengths)

6. To avoid the accident I _____
   _____

7. I _____ move my car before the officer arrived.
   (did/did not)

8. In my opinion, the accident was caused by _____
   _____
   _____

9. If you were injured, describe briefly _____
   _____

10. Note below and on reverse side names and addreses of your passengers and additional explanation of accident, if necessary.

11. Seat belts in use?   Yes _____   No _____

    _____
                        (Driver sign here)

    _____
                        (Address of Driver)

_____   Date: _____
(Reporting Officer)

Figure 2

During his interviews the Traffic Officer should avoid expressing opinions on the cause or any blame he may attach to it. Let those matters appear in his report.

Traffic Accident Reports are important. The liability of parties involved eventually may have to be established in court. The investigating officer may be called to testify to his findings at the scene of the accident. With court calendars as crowded as they now are, it may take two or three years before a civil trial will be held. Unless the officer thoroughly investigated and recorded the results of his investigation, he cannot adequately serve the involved parties. A good Traffic Accident Report often will determine whether or not the injured party will receive a just settlement.

Thorough and complete Traffic Accident Reports also may reveal traffic engineering problems at the accident site or the need for special enforcement directed at a particular violation that triggered the accident.

All Traffic Accident Reports should be reviewed by traffic supervisors with appropriate critiques to the officers who prepared them.

Parking

Campus parking poses its own share of problems for the security staff. Everyone wants to park very near his work area; everyone wants to park quickly on the way in, to leave quickly when he wishes; everyone wants his vehicle and its contents protected from theft, vandalism, damage.

Good parking plans are based on sound parking studies.

Parking inventories, for example, are useful in determining the geographical distribution of parking areas on campus, the capacity of each parking area, their physical condition, the degree of security risk they pose for users, and their normal use.

Like origin-destination or speed-delay studies, parking studies can be done by personal interview or by questionnaires. A questionnaire for this purpose might be in the form suggested in Figure 3. The study can be done at the scene when the driver parks his vehicle or by placing the questionnaire under the vehicle's windshield wiper blade and requesting that it be returned to a convenient point on campus. If it is to be mailed, it should be self-addressed with postage paid. The questionnaire should clearly identify the lot/space being used. It can be used in conjunction with origin-destination studies.

PARKING QUESTIONNAIRE

NOTE: Please complete and return as follows: _____

_____

Date: _____

Parking Area: _____

1. Status: Staff _____   Student _____   Visitor _____   Vendor _____

2. What was origin of your trip? _____
   Off-campus: Entered gate #:   _____
   On-Campus: Departed: _____

3. What building/activity is destination? _____

4. What is purpose of your trip?

   _____ Employment _____ Class _____ Delivery

   _____ Recreation

   _____ Use school facility: _____

   _____ Other: _____

5. How far is your parking space from your destination? _____

6. Parking times: In   _____ a.m./p.m.

   Out   _____ a.m./p.m.

7. Type vehicle: ____ Private   ____ School-owned

   ____ Commercial   ____ Other: _____

   ____ Car   ____ Van   ____ Truck

   ____ Motorcycle   ____ Bicycle   ____ Other: _____

8. How often do you use this parking area? _____

9. Did you encounter problems entering/leaving the lot: _____

   _____

10. Did you find space available? If not, how long were you delayed? _____

   _____

11. Suggestions (if any) for improvement of campus parking: _____

   _____

Figure 3

However the parking study is conducted, it should identify such significant parking problems as:

- Excessive, illegal, or overtime parking.
- Excessive cruising to locate a parking place.
- Excessive congestion and interference with traffic flow due to vehicles attempting to find parking spaces.
- Parking in excess of 400 feet from desired location.

Many schools have found that paid parking offers a good source of revenue; in some cases enough to cover the entire operating cost of the entire Security Department.

There are problems, however, to paid parking. School administrators should be aware of them.

If a person pays a parking fee, he has the right to insist upon:

- Adequate parking space, readily available.
- Parking space equal in size and quality to that offered all other patrons.
- Fair, impartial, prompt, and dependable enforcement of parking rules (*e.g.*, all pay the same fee, all park as directed, etc.).
- Adequate security for himself and for his vehicle.

Another problem is in collection of parking fees. Possible approaches include:

- Registration at the school includes a parking fee (this leads to administrative problems in processing applications, to include handling of applications received at other times and payment by staff and faculty members).
- Attendant-monitored parking (good but costly in terms of employment costs for the attendant and less expensive use of a contract guard often does not provide a competent attendant).
- Meter parking (see below).
- Commercial automatic gate controls (see below), perhaps used in combination with the paid attendant.

Many schools find the simplest solution is meter parking. The major problems involved, however, are maintenance of the meters and collection of fees. Meters can be easily damaged by improperly operated vehicles, by weather, by counterfeit coins, by vandalism, or by theft. And, for meter parking to be effective, the lot must be constantly patroled with citations promptly issued for overtime parking. Unless this is enforced, the parking plan will not work.

Problems in automatic gate controls include the following:

- Equipment requires frequent maintenance.
- Gates are subject to damage by vehicles, vehicles by gates.
- Winter weather affects gate operations.
- Card key controls (use of a special ID card for certain authorized holders rather than payment of a fee) can be ineffective as cards are lost or loaned.
- Gate area may require fencing or similar barricade to prevent drivers from by-passing the gates altogether.
- Patrols will find other ways to by-pass gate controls.

The fee collection procedure is the school's choice and often depends upon the size, location and type parking facilities available. Any of the above approaches can be made to work, but the administrator should keep in mind that each has its problems. Paid parking is no gift of the gods; it poses problems for the Security Director, and he must earn the revenue it offers.

The following physical features should be considered for the parking areas themselves:

- Lighting should be high-mast, clustered reflector type, each containing easily replaced, long-life lamps which are protected by mesh screen or resistant glass to reduce vandalism. For a very large lot, as in football parking areas, a solution which has worked involves use of six 1,000 watt mercury or metal-halide reflectorized lamps on a single high mast. It will adequately light an area equivalent to four football fields and replace the need for many lamp posts.
- There should be separate entrances and exits.
- Signage at the entrance of the lot should include: "Authorized vehicles only." Other signs should indicate other restrictions on use of the lot.
- Space should be reserved (and marked) for handicapped persons. Other use of reserved spaces should be extremely limited, *e.g.,* school officials only and visitors. When reserved, they must be properly marked and the use enforced.
- Space for individual vehicles should be clearly marked; angle parking is much better than 90° parking.
- The lots must be kept relatively free of snow during winter months; this requires plow scheduling in coordination with lot use.
- Trees, shrubs, or other obstructions to driving and to security observation must be kept free of the lot. If they exist on islands within the lot, keep them pruned.

The main problem in campus parking, as in traffic control, is enforcement of parking regulations. The school must establish a workable procedure for collecting assessed fees. This is a very sensitive area. Often faculty members, for example, will refuse to pay parking fees or assessed fines for parking violations. The school administration must support the Security Director in this, as in all other matters. They establish the rules; he enforces them. When individuals transgress, they must pay for it. Regardless of their position at the school or any other factor. All are treated alike.

Several collection plans are possible. Some schools have parking tickets returnable to the local community (usually more apt to be enforced); some have them processed as any other school fee. If the individual wishes to be part of the school community, he must honor that assessment. However the fee is collected, its amount is important. If the fine is too low ($1 − $2), it will be ignored. It should start at $3 for the first violation; $5 for the second violation; $10 for the third violation.

Sometimes the only effective way to handle those who ignore school parking regulations is to tow away the offending vehicle. If a vehicle seems abandoned and cannot be readily identified; if the operator is known to be an habitual violator of parking regulations and has not responded to citations; if the vehicle has no school decal and can't be identified, it may have to be towed away.

Towing is a last resort, however, and should be used only for flagrant violations. Before using the technique the first time, the school's legal staff should check local ordinances and be certain of insurance/liability considerations involved in any damage from the towing or storage.

If the school hires a professional tower, there must be a written agreement on such things as:

- Who will authorize a vehicle to be towed.
- Where and how towed vehicles will be stored.
- How stored vehicles will be released from impoundment.
- How each vehicle will be handled during towing.
- Charges per vehicle and how collected.
- The school's Security Department may require a vehicle to be towed, from time to time, without charge.

Towing should be done only with proper equipment which is operated by an individual trained, licensed, and authorized by the school to do that task.

Whenever a vehicle is towed from a campus parking area, a Security Officer should be present to witness the towing and to issue a traffic citation (affix to windshield of towed vehicle). He also should conduct a spot check of the license plate number with his Security Operations Center and with local police before the towing takes place. In his written report of the incident he should note any damage to the vehicle prior to towing.

A final consideration for campus parking involves the temporary parking incident to campus athletic or other special events. The following points are suggested for special events parking:

- As far in advance as possible, determine the date, time, exact locations which will figure in the special event.
- Estimate the turnover rate of vehicles which will be using the area.
- Survey the parking area; mark off areas to be avoided (pot holes, rocky areas, etc.); insure that surface will support anticipated traffic.
- Mark parking areas with appropriate signs or tape.
- Establish clearly defined entrances and exits.
- Establish clearly defined one-way traffic patterns.
- Establish type parking (head-in or 45° angle) and, where possible, mark parking areas.
- Establish and mark areas to be set aside for VIPs, emergency vehicles, etc.
- On-street parking: parallel rather than angle; no parking on roadway unless absolutely necessary; no parking near intersections (50 feet from crosswalks, 30 feet from stop signs).
- If angle parking is used, circular pattern must be one-way and clearly marked.
- Be certain that pedestrian flow does not conflict with main flow of vehicles.
- Assure that parking areas have reservoir spaces inside the exit area (30-40 feet) in which no parking is permitted. This will allow for uncongested buildup of vehicles leaving the lot.
- Review the availability of adequate Traffic Officer force to be on hand, assignments in special events area, availability of emergency back-up security force.
- If possible, have a towing vehicle available to clear the roadway during emergencies.

Proper control of campus traffic and parking involves serious research and planning with tight control over implementation of the resulting traffic and parking plans; a continual reassessment of their operational flexibility and adaptability in meeting changing conditions; strong support by school authorities to the Security Director responsible for this very sensitive area of campus security operations.

CHAPTER 8

RECORDS AND REPORTS

KEY TO THE successful operation of any security staff are records and reports. Security Officers, investigators, Traffic Officers, and supervisors must report significant things that happen as they perform their duties.

They must report them clearly, concisely, completely, and accurately so their supervisors can understand and put their input to use.

It also is important that report writers receive some feedback on their input. Does the report say what the writer wants it to say? Does it cover the who, what, when, where, and why (as appropriate) of an incident? Does it label hearsay information? Does it leave any questions unanswered? Is appropriate documentation attached? Are individuals and incidents cited in the report clearly and completely identified? Does the report reveal weaknesses in the reporter's performance of his duties (observation skills, people skills, knowledge of his job, ability to organize a report, etc.)?

As early as possible, the report writer should be critiqued on his observation techniques and on his report writing skill. Regardless of the quality of his work, the report writer should know that someone has read, understood, and considered his work. Otherwise, he is not going to be as enthusiastic or as careful about it the next time.

The Security Operations Center should be adequately stocked with all necessary logs and other report forms. Copy equipment should be available for additional copies.

FILES

Reference files containing technical references and necessary school and department publications should be available to the security staff.

They must provide information that is current and easily retrievable. Items used more frequently (*e.g.,* SOPs for departmental operations) might be available in several copies, depending on the size of the Security Department, for simultaneous use by several persons.

Incident files should be cross-referenced for research and investigative purposes. For example, 3x5 card files cross-referenced for three categories: locations of incidents, type incidents, and names of individuals involved. Such a system quickly highlights problem areas, type problems, and possible suspects for investigations.

Security Department files are sensitive and must be treated that way. Their control, maintenance, and use should be one person's responsibility. Individuals should not be allowed to remove material from them without a supervisor's approval with removal receipted until the material is returned. Files must be secured whenever their attendant is not on the job. If a large department has its files on computer tape, a back-up file should be maintained elsewhere to preclude their accidental loss or destruction.

A formally written and enforced policy should exist for automatic destruction of file material at the proper time. Most material, unless it involves an active investigation, probably should not be retained more than a year. Let the Security Director review and establish a destruction cutoff for each category, or a period in which they enter a dead file for a certain additional period and then are destroyed.

REPORTS

Security Operations Center Log

Each shift of the Security Operations Center should maintain an operational log documenting reported activities and incidents and individuals involved. Unless there is a separate sign-in and sign-out sheet for all security personnel, their names and times should be entered on the Security Operations Center Log. This is for historical and administrative purposes.

At the end of a shift, reports of Security Officers and their supervisors should be attached to the Security Operations Center Log for that shift, or otherwise be made available for review by the Security Director. He should review them each day, initialling to indicate that review and taking the follow-up action they may suggest to him. His review means that he intends to know all that is happening among his staff and on his campus.

Security Officer's Report

Each Security Officer must complete and submit a daily activity report for each shift he works. The report should contain the following information:

- Name.
- Date.
- Post Location.
- Supervisor.
- Security Officer relieved; time of relief.
- Relieving Security Officer; time of relief.
- Receipt of equipment, logs, etc.; delivery to relieving officer.
- Any noted fire, safety, or security hazards.
- Chronological account of Security Officer's location and activity during the entire shift.
- Summary of any unusual activity or incidents witnessed by that Security Officer or reported to him. Entries should be cross-referenced to the appropriate Incident Report more fully describing the reported incident and to any further documentation about it. All individuals involved in the incident or having knowledge of it should be fully identified. Some weeks from now the incident may assume considerable significance. Unless the Security Officer has recorded details in his report (and on an Incident Report), he may not be able to recall them. Certainly his testimony in a court of law, if taken from memory rather than from detailed notes taken at the time of the incident and noted on his shift report, would not be as strong.

Military time entries, rather than "A.M." or "P.M." are preferable. If the writer forgets to cite the "A.M." or "P.M.," confusion is likely; not so with military time systems.

Daily activity reports should be opened as soon as possible after the Security Officer signs in for duty and be completed throughout the shift (*e.g.,* as soon as possible after completing a patrol round, he should enter the time and route of that patrol and any unusual incidents noted). The report then will be current at all times and will not require much time to complete at the end of his shift. He must complete, sign, and submit his report to his supervisor before completing the shift.

Completed daily activity reports must be reviewed by the Security Officer's immediate supervisor before that supervisor completes his tour of duty. Delay may cause loss of specific information valuable to the Security Director.

Senior Security Officer's Report

This report is similar in format and intent to those of the individual Security Officer. The scope, however, is broader. This is a supervisor's report. He should record all assigned Security Officers and their assigned posts. The Senior Security Officer's Report should cite any problems that occurred on his shift (personnel late or absent, equipment malfunctions, etc.).

The Senior Security Officer must account for his supervisory activity during the shift (*e.g.,* "0400-0450: Inspected gate house; 0610-0645: Walked clock tour with S/O Jones") and for any absences from his post. He must cross-reference his report to Incident Reports he received during his shift and, if appropriate, indicate results of follow-up actions or investigations he took because of those Incident Reports.

He must review reports of all personnel assigned him, critiquing their performance of duty and their report writing (as appropriate), praising good work and correcting poor effort. He also must assure that outgoing Security Officers, besides completing their own reports, brief their replacements.

The Security Director should review his Senior Security Officers' reports. Every shift, every day, going further to check an individual Security Officer's activity report, if necessary, to clarify a question. It is the way Security Officers who patrol the campus communicate directly with him.

Incident Reports

All incidents of an unusual or questionable nature should be reported to the Security Operations Center. That can be done by anyone, whether he is a member of the security staff or not, who witnesses or hears of the incident reported. To get the information first-hand, the Security Office receiving a verbal report of such an incident should have the individual prepare an Incident Report. If this is not possible, the Security Officer should summarize his interview with the reporting individual and fully identify him for further interview if necessary. Hearsay (second hand) information, rumors, or opinions must be clearly labeled.

The important thing, whether the incident is an attempted robbery or a fall on the steps of the library, is that the Security Director be informed of it as soon as possible so that he can take necessary follow-up action. Incident Reports are very important tools. Individuals should be encouraged to use them to report anything they believe worth bringing to the attention of the Security Director. Security staff members *must* complete them for incidents they witness or have reported to them.

An example of a possibly innocuous incident which nevertheless should be reported on an Incident Report, would be Security Officers' stopping a motor vehicle, containing two youths, which they have observed circling a campus facility during the early morning hours. The Security Officers verify the identity of the individuals involved and of their vehicles. No evidence of illegal or improper action is noted, however, so the vehicle is released. Later, a series of thefts are discovered. They occurred in that facility during that time period. The Incident Report leads to the individuals questioned on campus that night and to the recovery of stolen material.

A wide variety of Incident Report forms (by whatever name they may be known) are available (local police, other schools, etc.) to the campus Security Director. He should use them and his own imagination to develop the form most suited to his own need. It might include the following information:

- Type incident (robbery, complaint, accident, etc.).
- Date and time reported.
- Individuals involved (completed identification).
- To whom reported; when reported.
- Person preparing the Incident Report; date.
- Date, time, and location of the incident.
- Narrative. Leave plenty of space, with instructions to complete on the reverse of the form if necessary. The who, what, when, where, why, and how (as appropriate) as known of the incident. It should include details of the damage or extent of injury (as appropriate), to include a complete description of any property involved (make, age, extent of damage, etc.).
- Other witnesses: name, addresses.

The form should include room at the bottom for the security staff supervisor who reviews the report to indicate action he recommends or has taken concerning it, and space for comment by the Security Director.

Referral Report

The Security Director may wish to refer information his department has developed to school authorities.

The Director should review and personally sign such referrals. It keeps him abreast of his department's actions affecting the school community, gives him an opportunity to review the matter before committing his department to it, and reminds staff and faculty members that he is personally responsible for his department's actions.

If a specific form is not felt necessary for such referrals, there should be an entry on the original investigative report or similar document leading to that referral that indicates that the information and appropriate recommendations have been referred to a specific individual in the school's administration.

Receiving Log

Often the campus Security Department will be required to receive packages or registered letters received at the school during non-duty hours. It should maintain a record of this material, from the time it is received until it is passed to the proper department.

The receiving log should cite:

- Dates.
- Times and department to which it was delivered.
- Individuals involved in the two transactions.
- Type material (letter, package).
- Senders/addressees.

Completed receiving logs should be certified by the Security Officer responsible for them and submitted to his security supervisor at the end of the shift.

Visitor Log

If visitor logs are used to control access and to identify visitors to more sensitive campus facilities (*e.g.*, computer centers, research facilities, etc.), they should cite:

- Dates.
- Activity location.
- Times.
- Visitors' names and companies.
- Persons or departments visited.
- Receptionist responsible for the log.

The receptionist should verify entries as they are made, before allowing the visitor to leave the reception area. This discourages fraudulent, incomplete, or improper entries. If escorts are assigned for visitors, their names should be noted in the visitor log.

Completed visitor logs should be certified by the individual receptionist responsible for entries on them and submitted to the security supervisor at the end of the shift.

TRAFFIC RELATED REPORTS

Much of the school's police activities deal with traffic matters. They require several special report forms, such as traffic citations, traffic accident reports, or vehicle registration forms.

Traffic Citations

Traffic citation forms were discussed in an earlier chapter. A variety of forms are available for this purpose. Some schools have found it helpful to develop different forms for moving and parking violations.

Essential to the form, however it is developed, are:

- Identification of the violator and his vehicle.
- The time and location of the violation.
- Specifically how the law (or school regulations) was violated.
- The identification number of the Security Officer/Traffic Officer witnessing the violation.
- The date of the citation.
- Specific action required of the person charged with the violation: payment of fines; location, date, and time of a required court appearance; check block for voluntary appearance if appearance not otherwise required.

Other data that might be included (particularly for more serious infractions) are:

- Vehicle description (make, color, body type).
- Driver's license number (if available).
- Driver's name and address (if available) and status at school: staff, faculty, student, visitor.

Traffic Accident Reports

Traffic accident report forms are important to properly record details of traffic accidents. In addition, by their format they lead the officer through the proper steps in conducting an investigation of that accident.

The liability of parties to the traffic accident, as well as the responsibility (if any) of the school, may have to be determined in court. That may take place months after the accident occurred. Unless the investigating officer has clear notes of his investigation, summarized on the accident report form, he probably will not be able to recall details concerning it. The traffic accident report is a legal and valid tool for him to use any time he must testify about the incident.

A good traffic accident report also may point out the need for improved safety features at the accident site, something of considerable value to the school.

A variety of traffic accident report forms are available through state and local law enforcement agencies and insurance firms, as well as from other schools. The Security Director should consider the variations present in this particular area and take the best of each to design the form most suitable for his own department.

Vehicle Registration Forms

Again, such a variety of forms are available for registration of motor vehicles, motorcycles, and bicycles that it is best for the school's Security Director to gather a few of the best available to him then develop the form most suitable for his own purposes.

It may seem that the various reports, logs, and other forms recommended here are time-consuming and not worth the effort they involve. If the school means to have a good security program, that is not true. Particularly if the program involves having a security staff to carry it out. The effort involved in maintaining all the logs, reports, and other forms is paid in full by a single successful investigation, by the savings from a civil suit that never takes place, by the supervisors' use of the various reports to develop the job skills of individual security staff members, by the improvement in operational techniques which have been made apparent through these records.

Once an individual becomes accustomed to the various reports and forms, he can complete entries with little difficulty in little time. It becomes second nature. He does not mind doing it. And there always are slack periods when there is ample time to keep the logs and reports current.

Accurate, complete, neat, professional reports and logs are extremely important in supporting investigations and subsequent legal actions. These attributes lend credibility to the department whose staff prepared them. Days, weeks, months after an event occurred, those individuals may be required to testify about entries contained in their logs or reports. Then it is perfectly appropriate for them to refer to these documents or to their notebooks during their court testimony.

The Security Director must insure that the work of his staff is thoroughly and professionally documented. Otherwise, much of their work may be of limited value.

CHAPTER 9

SECURITY SURVEYS

THE FIRST STEP to improving a school's security is to assess its present vulnerabilities, likely threats to those vulnerable areas, and the measures being taken to counter those threats.

A security survey identifies critical areas or functions within the school and safety or security hazards to them. It weighs those factors and recommends viable changes or additional measures to lessen those hazards.

The security survey should not be conducted as a clandestine affair, nor in a vacuum. It should be done openly, welcoming input by everyone concerned. The Security Director may not agree with the discrepancies reported through the survey, nor with the recommendations for correcting them, but he can't go wrong by listening and appreciating the interest and effort of the individuals making the report.

The survey should be as complete and as detailed as time and resources permit. It should include all the school's facilities and activities, day and night, weekends and holidays. It should cover areas outside the school's boundaries as well as those within it.

It could include many topics:

- General information on the school's physical plant: location, size, area and building use, sensitive areas and sensitive operations.
- Information on staff and faculty (numbers and responsibilities); students (numbers and activities).
- Identification of critical areas/equipment.
- Identification of vulnerabilities.
- Physical security/public safety requirements.
- Fencing (perimeter, swimming or other risk areas, etc.).
- Interior gates and doors.
- Lighting.

- Power and emergency power systems.
- Locks and padlocks.
- Key and combinations control.
- Signage.
- Marking of school property.
- Alarm systems.
- Closed circuit television (CCTV) protection.
- Fire alarm and fire fighting equipment/systems.
- Access controls (computer areas, residence halls, etc.).
- Shipping and receiving.
- Purchasing and contracting.
- Property control.
- Vehicle maintenance.
- Plant and equipment maintenance.
- Computer security.
- Security of residence halls.
- Security of museums and galleries.
- Security of cash and other valuables.
- Retail security (gift shops, book stores, etc.).
- Housekeeping services; garbage, trash, salvage.
- Mail and parcel post facilities.
- Dining facilities.
- Medical facilities; emergency services.
- Vending services.
- Communications.
- Employee background investigations.
- Campus investigations.
- Campus safety programs.
- Security and safety training and awareness.
- Liaison with local police and other agencies.
- Record of theft, vandalism, sabotage, misuse of school property.
- Emergency planning.

The Security Director must analyze security survey reports and, applying his experience and judgment, prepare or approve recommendations to correct weaknesses and discrepancies cited.

Because of economic or time restrictions, only a few corrective actions may be possible. Even then considerable time probably will elapse before permanent systems to correct discrepancies can be operational. During the interim, the Security Director must use imagination (signage, decoy cameras, lighting, adjustment of guard rounds, etc.) to lessen risks.

Department heads and other supervisors must be directed to correct security or safety weaknesses within their own areas of responsibility. They should be given a time limit to effect those corrections. The Security Director then should follow up to see that it has been done.

Even though some security and safety weaknesses cannot be immediately corrected, the Security Director's search for their existence is a form of showing the security flag. It also shows the school community that he is concerned about protecting them and about protecting school property.

The Security Director must do certain portions of the school's security survey himself. Portions may require his own expertise and experience; portions may involve particularly sensitive individuals and sensitive subjects.

Security surveys, however, take a lot of time. The Security Director has other responsibilities. Besides, working alone he may miss things. He can use responsible and knowledgeable assistants to do the leg work. He can evaluate their findings and follow up on them if necessary.

Members of the security staff probably will know more about day by day operations of the school, at the grass roots' level where problems are most likely to occur, than would that Security Director. Besides, participating in the facility's security survey is an excellent training device for key security staff members. It teaches them and it motivates them.

For his survey teams, the Security Director should select the most experienced, most mature, most capable individuals. They must be able to get along well with employees and supervisors of other departments. No one knows a particular department or activity better than those who work in it every day. They probably will have ideas worth considering. Again, the bonus is that involving employees gives them a sense of belonging and a sense that the school cares about them and the work they do.

The Security Director should personally brief and debrief survey team members. He should go over each report with the individual who prepared it. He then will understand exactly what the surveyor noted and can send that individual back for further information if necessary. The surveyor will profit by that debriefing and feel that his work is appreciated. He need not be really expert in safety and security measures. It helps, of course, but is not critical. He will have a checklist to help him and, if he is properly selected, the common sense to apply it properly.

The checklists, prepared and provided by the Security Director, will focus team members' thinking on areas his experience indicates should be covered and suggest minimum requirements adequate for safety and security coverage. The checklists, however, cannot be all inclusive and must encourage the individual surveyor to exploit his own thinking and initiative.

The remainder of this chapter suggests some areas and operations which should be examined. These lists are not all inclusive. They merely suggest. Consider them in terms of your own school's requirements, then prepare your own checklists.

Signs

Campus signs (*e.g.,* campus limits, restrictions, gate use, directions, parking, traffic control, identification of facilities, etc.) must be adequate in number, size and wording; appropriate for their intended purpose; properly placed; visible; and in good repair.

Fencing

If a barrier exists around any portion of the school's perimeter or an internal campus facility, cleared zones free of debris or other material or growth offering concealment and hindering observation should be established and maintained 15-20 yards (if possible) on both sides of the barrier.

Paths or roads coming close to the perimeter, makeshift ladders, or other means of crossing the barrier should be checked as they may indicate improper activity at a particular point on the barrier.

Security patrols periodically should check buffer zones for signs of intrusion, damage to the barrier, debris, or the need for additional lighting or additional security surveillance.

Lighting should be directed down and away from the barrier. It must not blind security surveillance and should be placed and aimed so that, if a light is inoperative, its area still will be lit by an adjacent light. The security surveyor should check all lighting at night, as well as during the day, to be sure that all lamps are operating. He must evaluate reporting and replacement procedures for defective lamps.

Gaps or other damage to fencing must be corrected. Fencing itself should be sturdy, well anchored, at least seven feet high. If strands of barbed wire are added, they should be angled outward. Unauthorized openings (damaged fences, soil erosion, exposed drainage ditches, etc.) must be corrected.

If buildings are part of the perimeter barrier, windows accessible from the ground must be secured or sealed and protected by grills, mesh, or special glass. Exterior doors or gates must provide as effective a barrier as the fence itself.

Area/Building Security

All areas immediately surrounding campus facilities should be clear of plantings, debris, or other obstacles more than two feet high which might offer concealment or hinder security observation.

Parking lots must be adequate in size and surface for their anticipated use and appropriately sited (near use areas, with direct traffic flow); supported by adequate and appropriate signage; protected by lighting, security observation, and fencing (if appropriate).

Buildings' doors and their locks should be appropriate to the area they protect (expensive lock on an easily forced door, etc.). Fire or safety codes must not be violated by any improperly locked doors. Exterior doors should be adequately lit for safety and for security monitoring.

The security surveyor should check all exterior windows (or interior doors and windows to sensitive areas), particularly those accessible from the ground or from other structures, to be sure they are adequately protected by mesh, safety glass, or security glazing. Appropriate windows and skylights or other entrances to the roof must be alarmed and denied easy access from the outside.

Alarm systems must be tested during every shift. All alarm and CCTV surveillance systems must have emergency power back-up and be not easily susceptible to damage.

The surveyor should thoroughly check all security patrols (routes, times, Special Orders, detex clocks, etc.) for adequate coverage.

Lighting

The security surveyor must assess lighting systems, paying particular attention to parking lots, sidewalks, residence halls, receiving docks, warehouses, and other particularly sensitive areas.

Light fixtures must be protected against vandalism and have emergency power back-up. Independently powered, portable spotlights should be available to the security force.

Lighting should be checked at night: proper aiming, brightness, overlapping beams, nonoperating equipment, and procedures for reporting and correcting defective equipment.

Key Control

Perhaps the greatest weakness to most physical security systems is inadequate key control — too many master keys floating among unauthorized individuals. Locks are not changed when an individual who has a key or combination no longer needs access. Issued keys are not collected when an individual leaves or changes assignment. Keys are issued as a convenience to many individuals when the availability of a very limited number of master keys, under the control of the security staff, would be as efficient and far more secure.

The surveyor must evaluate the school's policy on who is authorized to receive keys or combinations and, if an individual is not authorized or has no real need to have that key or combination in his possession to do his job, he should not have it. Let the security staff open the door for him. Authorization to issue keys should be centralized; the same with duplicating keys. There should be policy, established and enforced, to handle return of keys no longer needed, changing of combinations, and processing of lost keys.

Locks must be appropriate for their use. There is no point in using an expensive lock when the door itself can be removed by stretching the door frame, by removing hinges, or simply by splintering the door itself.

The surveyor should check use of the master key box in the Security Operations Center to insure that it adequately protects and accounts for all keys at all times.

Closed Circuit Television Surveillance

CCTV surveillance systems are very expensive, easily damaged, require frequent service and careful use. The surveyor should check all cameras and monitoring equipment and their use. Lighting (to include an emergency power back-up) must be adequate for the cameras used. Cameras should have tilt/pan/zoom capabilities, be properly aimed, be equipped with time lapse taping capabilities. They must be protected from vandalism, sabotage, theft.

Security Officers (specifically designated for that duty and rotated periodically throughout a shift) must be assigned to monitor surveillance cameras. Cameras and monitoring chairs must be placed so that the monitoring Security Officer will not be uncomfortable. A sequential check of all cameras at one screen is far preferable to his attempting to scan a dozen cameras at one time.

If the monitoring Security Officer is not properly trained, equipped, and motivated to monitor the information his camera provides, his is a wasted effort. If he has conflicting duties (answering the Security Operations Center telephone, receiving visitors, answering questions, etc.), he may not note what his cameras are trying to tell him. Then the system is wasted.

He must have someone available to respond to matters revealed by the surveillance cameras.

He must maintain surveillance logs. At the end of each shift, each log should be checked by the shift supervisor.

Entrances and Exits

The security surveyor must be alert for any signs of unauthorized entrances and exits (*e.g.,* gaps in perimeter fencing, jammed residence hall locks, opened fire doors, etc.).

Security Guard Force

A senior, highly experienced mature security staff supervisor should evaluate the campus security staffs' operations, particularly those of the security guard force.

He should review the contract with any commercial guard company serving the school to evaluate such items as: verifying proper charges for the company's services, accounting for times worked by each guard, the availability of equipment and uniforms called for by the contract, supervisory controls over guards' performance.

He should weigh the training of individual Security Officers, contract and proprietary.

He should check all Security Department equipment for proper maintenance and serviceability and for its accountability.

He should review all guard orders, guard rounds, emergency plans.

He should pay close attention to the Security Operations Center, nerve center of security operations on campus. Is it adequate in size, configuration, location, staffing? Is it properly equipped? Does it have emergency power back-up? Are its operational procedures effective, economical, based on common sense?

He should study the work of guard supervisors. If they do their job well, the staff will do good work. If they do not, the best of organizations and the best of plans won't work well at all.

He should review a number of Security Officers' daily activity reports, randomly selected from all shifts. Are they complete, neat, accurate (so far as he can determine)? Using selected shifts, he should spot check to be certain that all Security Officers are submitting reports. Have supervisors checked each report, and are their reviews worthwhile? Do the reports or the supervisors' critiques indicate a need for additional training or other measures?

Have reports, and appropriately cross-referenced Incident Reports or similar supporting information, been properly filed and secured for future reference?

Solely from a review of a Security Department's records and reports, the security surveyor can gain an excellent assessment of that department's operational efficiency and weaknesses.

Fire Safety

The surveyor must assess existing fire prevention and protection measures for adequacy, currency, completeness, and suitability. Are testing procedures adequate, and are they used? If the school has not had a fire drill in some time, now is an excellent time to have one. It may save lives.

Alarms, sprinkler systems, extinguishers, and hose systems should be checked for suitability, adequacy, and serviceability. Are fire fighting teams organized for each shift the school is in operation? Are they organized, trained, drilled, accounted for each day—or are they a fire fighting team in name only?

As he conducts his survey, the surveyor should be alert for fire hazards (debris, combustible material, defective wiring, circuit overloads, blocked sprinklers, blocked fire exits, etc.). He should have them corrected on the spot.

Purchasing Procedures

Purchasing procedures should be studied to identify favored vendors. If some are there, the purchasing agent should be required to explain each one.

Purchase orders should be printed on unalterable paper and be numbered, properly secured before use, and properly filed after use. Periodically, a responsible school official should check purchasing and receiving documents for accuracy.

The surveyor should insure that purchased materal is adequately and safely stored to preclude its being damaged, misused, or stolen. Is it entered on the inventory? He should review inventory procedures and personally spot-check a number of items as they are received.

Shipping & Receiving

If the school is involved in shipping procedures (book sales, gift shop items, etc.), the surveyor should review sales and order filling and mailing procedures. He should spot-check some addresses and some packages ready for mailing.

His survey should assess:

- Shipping and receiving procedures.
- Shipping and receiving areas' physical organization, lighting, alarm systems, security guard coverage.
- Housekeeping.
- Presence of trash or trash disposal containers near shipping and receiving areas.
- Inventory procedures.
- Supervisors' control of employees.
- Control of delivery drivers (held in outer dock area, controlled if allowed to move into the receiving area).
- Package control among employees.

Property Control

The surveyor should assess such property control measures as:

- Assignment of responsibility to specific individuals for all school property.
- Property books; issue and turn in procedures; inventories; accounting for losses.
- Storage procedures.
- Marking of school property.
- Package control.
- Maintenance procedures.

- Waste, salvage, disposal procedures.
- Record of theft and vandalism of school property.

Vehicles

School vehicles must be properly protected and properly used. When not in use, they should be kept in a secure parking or storage area. When needed, they should be dispatched (in writing on a vehicle dispatch log) to a specific individual (authorized and licensed to operate that equipment), for a specific purpose.

All vehicles should be properly maintained by the driver and by the school facility or contract agency responsible for such maintenance. A complete record of maintenance performed on each piece of equipment should be maintained for one year after that service is performed.

Retail Operations

The security surveyor must assess the physical layout and operational procedures of all campus retail facilities (book stores, clothing stores, gift shops, cafeterias, etc.). He might look for the following:

- Are service entrances locked from the inside when they are not in use?
- Are aisles and counters neatly arranged and clear for observation by sales and security staffs?
- Do any of the sales displays lend itself to theft?
- Are critical or sensitive items (watches, cameras, calculators, etc.) stored in separate, secured storage areas?
- If the activity's safe is small, is it secured to the premises by chain and padlock? Is a strong light left on over the safe when it is not in use? Has the safe's combination been changed in the last year or since someone having knowledge of it has left the facility's staff?
- Does the manager, at the end of a business day, inspect the premises to insure that all windows, doors, safes, etc. are closed and, where appropriate, locked; that no person is hidden on the premises; that intrusion alarms have been activated?
- If the facility is equipped with a silent holdup alarm, is it monitored in the Security Operations Center?
- Are personnel familiar with school policy regarding shoplifters and trained in handling them?
- What action is taken when high value items are discovered missing? Is the action appropriate?

- When it is necessary for any employee (or anyone else not on the retail facility's staff) to enter stock rooms or other areas where goods are not secured, are they escorted or monitored by a member of the sales staff?
- Are garbage and trash bins inspected for hidden or inadvertently discarded merchandise or other valuable items?
- Are employees required to leave all packages and purses in a designated secured area? Is a package inspection system in operation? Is it effective?
- Are employees required to check with their supervisors before leaving the facility?
- How are sales recorded? Is the system adequate and satisfactory?
- What procedures are followed for returned merchandise? Credit transactions? How is returned material returned to stocks?
- Are excessive quantities of money removed from cash drawers during the day?
- Are surprise cash counts conducted?
- Who audits cash register tapes? Is the system satisfactory?
- Are cash register drawers locked when not in use?
- Are check cashing customers properly identified? Does the facility accept post-dated checks?
- Is adequate control maintained over material taken into fitting rooms?
- What procedure is followed when inventory shortages are discovered?

Libraries

The surveyor should thoroughly check library facilities for fire protection programs:

- Proper extinguishers, adequate quantity, appropriately located and clearly marked for location and procedure for use.
- Halon or other extinguishing systems in good working order.
- Heat and smoke detector alarms in good working order.
- Library as free of debris, clutter, flammable material as possible.
- Exits, fire doors, stairwells clearly marked and unobstructed.
- Fire teams in being, properly equipped, trained and drilled.
- Well written procedures for fire alarms, drills, and evacuation procedures.

The book check system at the circulation desk should minimize confusion and delay in returning or borrowing books yet offer the best chance at detecting anyone's attempting to bypass checkout procedures.

Library supervisors must issue keys only to those who are *authorized* to have a personal key and who really *need* that key to do their jobs. Master keys should be limited and retained in security control. Unused keys must be secured and accounted for at all times.

The security surveyor should look for close and harmonious working relationships between the library and security staffs. Each must have an understanding of the other's duties, responsibilities, and operational procedures.

Security staff members stationed at library entrances and exits should be assessed on their effectiveness in deterring theft and in performing their public relations functions.

Access to stack areas, rare book rooms, audio-visual areas, and other sensitive areas, must be restricted to staff and patrons who really need to use those facilities. Library supervisors must identify users and monitor their activities in the secured areas. Extensive use of fascimile copies of more sensitive documents, rather than the original manuscripts, should be practiced.

Stack areas, reading rooms, and reference areas should be physically organized to permit as much surveillance of patrons as possible.

Alarms, closed circuit television surveillance, and other technical security measures should be audited by the security surveyor to insure that each device is appropriate, properly located, serviceable, and properly monitored by a staff member.

There should be a low temperature alarm to preclude frozen pipes during heat failures. Manuscripts and records in areas near water pipes may have to be removed if there is danger of flooding.

Whenever the library is open to the public, staff members periodically should move through all public use areas.

Cleaning Staff

Whether in office areas or residence halls, cleaning personnel often become involved in campus theft, vandalism, and pilferage problems.

They usually have complete access to all facilities and to all areas within each facility (not a good idea at all); they work without close supervision (often necessary); they are in sensitive areas at night or at other times when there are no other personnel in the areas; they carry trash or other large bundles from facilities without inspection or question (again, not a good idea at all); their employees usually are marginally paid and have not been investigated (particularly concerning their work history or past involvement with police authorities). In sum, they may be more tempted to steal or misuse school property than most other employees.

The security surveyor should study the custodial staffs' work throughout the school. Some concerns might be:

- Do background investigations reveal previous arrests, previous employment dismissals, or questionable work or personal habits?
- If employees are contracted, are they bonded? How closely are they supervised by the contracting company? Is there excessive turnover (new personnel) within the contracted cleaning staff?
- Does the cleaning staff have keys? Are they issued each day by the security guard then recovered? Does the guard open and lock certain areas for them?
- Is their work monitored by a security guard or by closed circuit television? How effective is that monitoring?
- Are the contents of service containers spot-checked by a security guard for pilferage?
- If the cleaning staff is too large for an individual to be readily identified by guards or other employees, are they uniformed or wear distinctive identification badges?
- If cleaning staff members who are school employees are caught stealing, are they immediately discharged? If contract employees, are they discharged or transferred from school duties? Are any prosecuted in court?

These lists obviously are incomplete in topic areas and in the thoroughness with which they cover specific areas. They don't begin to include all possible concerns for security and safety on a school campus.

Others are entirely too detailed for most school's operations. No matter. They are food for thought which should be adapted to a particular school's requirements.

The Security Director should use them to develop more extensive and more appropriate checklists for his own campus.

No matter how extensive his own lists (guidance to security surveyors), however, his security surveyors are going to find additional concerns they must investigate.

Security surveys involve a lot of hard work: much walking, talking, looking, listening, writing, thinking. They take time. Individuals involved in them should be carefully selected, properly briefed, and given adequate time to do their work properly. When they submit their reports, the Security Director must be willing to study those reports and to appreciate the effort they represent.

If all that is done, and if the Security Director will follow up on weaknesses discovered by the surveys to get those weaknesses corrected, the school will be safer and more secure, and the school community will know it.

CHAPTER 10

PROTECTION OF BUILDINGS
AND RELATED ACTIVITIES

THE BEST protection of campus buildings and their related activities begins when the school's Security Director can sit with campus architects and planners and provide specific input to planning for a new building or modifications to an existing one or to an activity to take place in that building. He has the professional expertise they need to build proper security systems into the physical layout and planned use of the facility. If that sort of ground floor coordination is not possible (*i.e.,* the building is there, activities already have been established in it, etc.), he must work with what he has to create a system to protect the building, school property within it, and the school community which will be using it.

His security "system" should include full use of his security staff (patrolling, monitoring security technical equipment, performing other human security services); the school staff which will occupy or work in the building; students and visitors; and as much technical security equipment (alarms, lighting, CCTV, locks, etc.) as is appropriate to that particular situation and available to him.

This chapter discusses elements of such a security system. A particular school may not require all the protective measures cited, or it may not be able to afford them, but they can be adapted to the Security Director's particular situation.

Incidentally, when planning security measures, school administrators would do well to ask students where they feel concerns. Most of the time, students will know for themselves areas where security should be of greatest concern. And addressing those areas will, in itself, ease those concerns.

Fencing, like other physical measures of the security system, is not foolproof. It will not stop a person who is determined to cross it. It may discourage or deter him, however, and it certainly can slow him down. If the fence also is protected by patrolling guards or by television surveillance, that may be enough.

There is legitimate use for fencing on campus:

- To physically and psychologically deter unauthorized or accidental entry (as in a swimming area).
- To enhance guard effectiveness.
- To direct and slow vehicular or pedestrian traffic.

To be effective, fencing must be well maintained. It must be kept free of debris, vegetation, or man-made objects that might damage the fence, provide means of circumventing it, or offer concealment against surveillance. It must be patrolled or under other visual surveillance.

Fencing should be chain link, of 9 gauge (or heavier) wire, at least 8" high, with a mesh not less than 2 inches square. The wire must be kept taut and firmly secured to steel posts anchored in concrete. The lowest strand of fence wire should come within 2" of solid ground, to ground level on sandy, easily eroded soil. Culverts or ditches passing under the fence, with openings greater than 96 square inches, should be covered by similar sized wire mesh. If the protected area requires additional protection, a top guard extending outward from the top of the fence at 45° for 2 feet, and consisting of four (4) strands of barbed wire, may be used.

Gates must be as strong as the fence itself, and both must contain appropriate signage calling attention to the purpose of the fence (identifying property, establishing trespassing restrictions, etc.) and providing directions to those who would cross it.

Lighting

Probably the best physical measure to protect a campus facility, aside from the presence of human beings, is light. Those who would steal or damage property prefer not to do it in a lighted environment.

Roadways and sidewalks should be well lit. Vehicular and pedestrian areas should have lighting of an intensity of 2 foot candles; unattended outdoor parking areas: 1 foot candles. Sodium vapor lamps (provide golden yellow light) probably are most effective for campus use.

Lamps should be directed down and away from the areas they protect and lighting patterns should overlap so that, if one lamp is inoperative, the adjacent lamp still will cover the area.

Wiring to outdoor lighting should be underground if possible. Lamps should be protected by impact-resistant glass or glazing or mesh.

Lighting systems must include emergency power back-up and/or battery power (indoor lamps). Portable spotlights should be available to the security staff.

Maintenance of lighting is critical. Burned out bulbs must be detected, reported, replaced without delay. Normally, this is where Security Officers on patrol can perform an additional function.

Defective lamps, reported on their shift reports, should be repaired without delay.

Closed Circuit Television

CCTV surveillance systems are expensive to install and to maintain. Cameras must be protected (spray paint, thrown objects, pellet guns, etc.) and require proper lighting, a taping capability, proper placing to be effective. They are effective, however, and greatly extend the security staff's area coverage.

Cameras should be monitored in the Security Operations Center with a Security Officer available to respond to emergencies. Monitoring Security Officers (women are best in that role) should be trained in use of the equipment and in proper surveillance techniques. Monitors should not work longer than an hour without taking a break and should not be required to perform other simultaneous functions (answering telephones, completing reports, etc.). Let them concentrate on their camera monitor. Incident logs, supporting the video tapes, must be maintained.

Emergency power must be available for the CCTV surveillance system during power outages.

Alarms

Vulnerable doors, windows, roof openings to sensitive areas should be secured by alarm systems. There are so many types available today that, in designing alarm protection for a facility, the Security Director should consult a reputable alarm company.

Local alarms (sound locally when triggered) have a particular purpose (primarily to frighten away an intruder) but are limited. If no one can hear them (perhaps 500 feet), their value is limited, or to a knowledgeable intruder, nullified. Most intrusion alarms should be tied to a central control station, whether it be in the school's Security Operations Center or with a commercial security company.

The Security Director should periodically audit all alarm systems installed and serviced by a commercial alarm company (*e.g.,* Wells Fargo, ADT, etc.). He should *test* to be sure that each alarm works properly and should do a *physical inventory* of every piece of equipment for which his school is charged. The audit is a time consuming practice, however, and he might use a responsible staff member to make it. It is a very valuable exercise. If it is a system which has been in place for some time, particularly with the same commercial company, chances are he'll find alarms that do not work, were never installed, were not removed despite a school purchase order directing its removal, were installed but subsequently removed without credit to the school's monthly billing. If so, he can realize significant future savings and possibly win appropriate refunds from the alarm company.

Alarms must be monitored with sensings noted in an alarms log and investigated by a Security Officer. Each alarm should be tested daily with results entered in the alarms log. Security supervisors should check alarms logs every shift and take appropriate actions, when necessary, to follow-up on reported incidents.

Locks

Locks also are not a positive bar to unauthorized entry. They only deter or slow that entry. But in that they are a necessary start.

Locks must be sturdy or they are wasted; similarly, a good lock is of no value if the door itself is of hollow construction or otherwise so weak that it can be easily forced or its frame easily stretched by jacks to allow the lock's strike to fall free.

Dead bolt locks are preferred as they are much harder to circumvent. Interchangeable lock cores also are preferred as they are easy to install and remove, less expensive, and can be keyed to a master key system. As with alarms, when selecting locks for your school, consult a reputable local locksmith.

The greatest security risk on a campus probably is inadequate or, more likely, non-existent control of keys and combinations. Every department head, professor, or lesser Indian, wishes and feels he should have a master key to his building. Never mind that a Security Officer may be available to open locks for authorized individuals or that staff and faculty members seldom are in the building during normal off-duty hours and are prone to forget or lose the keys anyway. It's much like having a key to the executive wash room. Security Directors must constantly protest issuance of keys for sensitive areas to individuals who really do not need them.

Master keys and combinations should be issued only to individuals who are *authorized* entry to the area at all times and who have a real *need* for their own keys. If a Security Officer is available, the better solution is to keep all keys (particularly masters) in a locked key cabinet in the Security Operations Center to be issued, as needed, to a Security Officer to open necessary doors then be returned to the key cabinet.

The key cabinet should be secured at all times. A key log should receipt keys taken and returned, and all keys should be inventoried at the start of a shift. Keys should not be loaned to individuals. Keys should be number-coded, rather than openly labeled for the locks they open.

The Security Operations Center should have a formal procedure for issuing keys, for investigating lost keys, for changing locks and combinations, and for recovering issued keys when they no longer are needed. Key control belongs with the campus Security Department, not with its housekeeping services. It can become a very costly and time-consuming operation, however, and should be alloted proper funds and clerical support.

Locks and combinations should be changed annually, when any individual who possessed that key or combination no longer needs it, or when an incident occurs which causes concern that the lock or combination has been compromised.

Campus Offices

The degree of security controls exercised over campus office areas depends upon the sensitivity of the building and its related services. A building used solely for classes might require only limited security services (protection of library or resource material, audio-visual equipment, copiers and other office equipment). Buildings having computer facilities, finance offices, student records, etc., might require considerably more attention.

All first floor windows to campus offices should be held to the minimum number necessary, should be protected by shatter-resistant glass/glazing, or by mesh; and should be locked and possibly alarmed.

All other openings into the building should be secured against unauthorized entry or damage by water, fumes, other substances.

Office equipment should be inventoried; assigned to individuals, where appropriate; and, where possible, be secured to desks. School property should be clearly identified and serial numbers recorded.

Housekeeping and maintenance personnel should not be issued keys if a Security Officer is available to open doors for them. Their work should be monitored, as much as possible, by the security staff and their equipment storage areas checked periodically for secreted items or flammable material. All office areas should be kept neat, clean, uncluttered. It helps prevent fires and thefts.

Residence Halls

Probably the most critical campus area to protect against theft, violent crimes (assaults, robberies, rape), and fire are residence halls. Their occupants often make the job more difficult than others.

Install excellent security in the entrance way and student residents will leave other entrances propped open, loan entrance keys or combinations, find other ways to circumvent lobby security. They find it more convenient and more neighborly to do it that way. No one can deny that it would be great if locks and security guards were never necessary or that the threat of personal harm does not exist. Unfortunately, however, human nature being as it is, threats do exist; protection is necessary.

Shrubbery around residence halls should be trimmed to within 2 feet of the ground to preclude their being used for concealment. Thorny shrubs are better, for the same reason.

Walkways should be kept clear of close shrubbery and be lit (2 foot candles intensity).

Exterior lighting (at least 2 foot candles intensity) should be directed down and away from the building. Entrance ways, lobbies, and fire escapes should be clearly lit and visible to observation from the street. Entrances should be limited to one, if possible, with additional panic bar-lock equipped fire exits. Entrances to residence halls should face toward the street. All to enhance observation of the building from the street. The objective is prevention of crimes, not apprehension of perpetrators after a crime has been committed. Surveillance deters crime.

Entrance doors should be sturdy, locked, alarmed. Remote controlled locks (electronically released strikes), cipher (number coded) locks, and card-key locks are good for this purpose. The problem, however, is that the locks will work only as well as the students let them work (propped open doors, duplicated keys, loaned keys, vandalism to locks, etc.).

Lobby mailboxes should cite the individual's last name and initials, not the complete name.

Alarms throughout the residence hall, particularly door or window alarms, are helpful but must be monitored at a central station with a Security Officer available to respond when they are triggered. They are expensive and are not effective for the school's perimeter security. Use them for more restricted, more sensitive areas.

Locks to individual rooms or apartments should have interchangeable cores for changing when a key is lost, when there has been unauthorized entry by a duplicate key, when the occupants change.

Peephole viewers and night chains add considerably to the actual security of a room and to the occupant's perception of that security.

Security patrols should be planned during the architectural stage (if possible) to tie-in with other security systems and to assure complete coverage. Detex watch key systems are fine to audit the Security Officer's mobile patrol but work best for a community college or other campus where activity and facilities are more concentrated.

The Security Director should consider enlisting student support in the form of student campus patrols, particularly of residence halls. Peers are better received than Security Officers and participants can develop an increased awareness of the need for good security and communicate that need to other students.

Housekeeping areas should be checked for fire hazards and possible stolen goods. Close liaison should exist between the security and the housekeeping staffs. Housekeepers are close to students and often will become aware of potential or existing problems which, when shared with the security staff, can be corrected. The housekeeping staff should be issued no more keys than absolutely necessary; no master keys if that can possibly be avoided.

The greatest danger to residents is fire. Students block fire escapes, tamper with extinguishers, clutter halls and rooms. Most are poor housekeepers and, left unchecked and uncorrected, will allow health and safety problems to develop in short order. Fire prevention measures are discussed in a separate chapter.

Food Services

Many schools contract with commercial firms to handle campus food services. If so, the campus security representative may have limited security responsibility over food service facilities and services. He may be limited, for example, to providing security patrols and to linking alarms to his Security Operations Center.

Always, however, he will have some responsibility for security of these sensitive areas. Even when an outside firm runs cafeterias, pizza parlors, vending areas, if there is theft, vandalism or other damage to property, the matter will spill over to the school, involving students and administrators.

The most vulnerable areas are theft of food in storage or theft of cash from food service revenues.

Most theft will be by dishonest employees and may include theft from cash registers, theft from food serving lines, or theft of food in storage.

Many security concepts suggested elsewhere in this text for campus retail facilities also apply to food services.

Receiving clerks must be trained and required to do accurate counts and to verify the type and condition of all food or related items received by the kitchen. Deliveries should be stopped at the entrance to the delivery area, with drivers not allowed to wander unsupervised through kitchen, storage or serving areas.

Periodically, inventories of goods received or stock inventories should be spot-checked by supervisors.

Stockrooms, freezers, and refrigerators should be locked at all times. They should be under the responsibility of one individual who exercises strict control of keys and who assures limited access to these areas. If there are windows, they should be barred.

If thefts persist from stock areas, it may be necessary to install intrusion alarms or CCTV surveillance.

Food service supervisors and security personnel periodically should check trash and garbage for concealed or inadvertently discarded valuables or silverware.

The Security Director should maintain close contact with the vendor who picks up trash and garbage to retrieve items the vendor recovers from the school.

The security staff should spot-check food service employees' packages for stolen food or other school property. If an individual is caught stealing school property, he should be dismissed.

The Security Director also should spot-check the school's purchasing procedures for food services, for improper favoritism to certain vendors or for vendors' kickbacks to food service managers or purchasing supervisors.

Computer Facilities

The risk to campus computer facilities is not as great as to similar commercial centers but remains great. Computer equipment is tempting to one who wishes it for his own use or for resale. There also are the risks of individuals' unauthorized use of the school's computer facilities to obtain information for personal use or gifted students' improperly using their mental talents and their school's computer facility to enter banks' electronic data systems to fraudulently divert great quantities of money or to make fraudulent credit card purchases.

Normally, the security staff must rely on the school's computer staff to handle security of software programs, codes, etc. The security staff, however, will establish, monitor, and perhaps administer the physical security of the facility itself.

The campus computer facility must be given all the exterior physical security measures already discussed: a sound structure in which to operate and all openings secured with protected doors and windows (guards, lighting, locks, alarms, CCTV).

Access should be limited to the minimum number of people neccessary and be controlled by:

- Access rosters.
- Identification of each person using the facility.
- Maintenance of logs identifying each user.

There should be strict Standing Operating Procedures specifying who can use the facility and its systems; which systems they can use; any restrictions on its use; how they are to use it. User logs should be established and maintained with periodic audit and verification by supervisors and the security staff.

Valuable tapes, disks, and other software should be stored in vaults when not in use, and there should be adequate and protected emergency back-up files located elsewhere for each critical item.

Access to terminals and data logs should be limited and strictly controlled. Supervisors periodically should change keys, passwords, and codes; verify the currency and validity of authorization rosters.

Security procedures should be established and enforced for receiving, handling, and storing all supplies and equipment.

All personnel habitually involved in the computer facility should be screened by a background investigation.

All areas of the computer facility must be kept neat, clean, and free of debris, flammables or other fire hazards. Fire detection and suppression systems (halon) must be in place and must be inspected and tested at least weekly with logs recording each check.

A strict fire prevention and fire fighting procedure should be written, taught to all appropriate staff members, and posted in the computer facility. It should include action to be taken when an alarm sounds, a fire occurs, or there is a power outage. All emergency systems periodically should be tested with results reviewed by the computer facility manager and the Security Director.

Medical Dispensaries

Campus medical dispensaries, unless they are part of a large university hospital, are small and handle a limited amount of drugs or other medications. If any are present, however, they must be protected.

Campus dispensaries also are vulnerable to theft of medical instruments, office equipment, or other college property.

Dispensaries should be equipped with sturdy doors and locks; barred windows; intrusion and fire detection alarms. Their interiors and exteriors should be well lit and, if necessary, be covered by CCTV surveillance.

Personnel working in the dispensary should have available an emergency power supply and emergency communication, including a hidden holdup alarm.

Drugs must be secured and dispensed as the law requires. Locked metal cabinets with strict key and access controls and alarms should be used to store drugs when they are not actively being dispensed. Stored drugs should be inventoried and receipted for at the start of each shift.

Procedures for after hours access to the dispensary, and to cabinets containing controlled substances, should be written and must be strictly enforced.

The Security Director should conduct periodic spotchecks, to include audits of drug inventories, to evaluate the dispensary's security systems. The spotcheck might include an audit of sample medications, provided the dispensary by drug firms, to assure that controlled substances have been picked up on the dispensary's inventory and are being properly safeguarded.

School Property

Typewriters, television sets, calculators, microscopes, copy machines and computers are only a few of the tempting targets among school property items.

Storage areas for sensitive items should be walled-in (avoiding false ceilings of acoustic tiling), well-lit, equipped with solid locks, alarms, possibly with CCTV surveillance. More valuable items, particularly small, easily concealed ones, should be protected in a separate chicken wire or wire mesh cage.

No item of school property should be transferred from the responsibility of one person to that of another without written authorization for the transfer and signed receipts documenting the change in responsibility. Master property files should be established and maintained to accurately and currently locate all items at all times.

School property should not be taken from the campus except for official school purposes, and the action must be completely documented.

There should be a property custodian with records completely identifying every item of school property of more than $50 value (name, serial number, description, separate accessories, color, size, model, etc.); its location; who is responsible for it. There should be an annual inventory of all school property to verify its presence and its condition. Periodic additional spot inventories also are helpful. Identification numbers, if not already permanently affixed to school property, should be accomplished by the property custodian.

The basic intent is to have a complete chain of control for all school property at all times.

Some department heads won't like the idea of signing for school property to be used by their department. Nor for being held personally responsible for its safety, accountability, and proper use. But, if they sign for it, there's a very strong likelihood they'll take far better care of it than if they had not signed for it. If the school administrator requires one to pay for an item lost through neglect, it will work wonders for a security system.

Theft of private property from individuals can be deterred through the physical security measures already discussed and through a good security education program. The Security Director also might make available an "Operation ID" program where, when requested, the Security Department etches the owner's name (electric marking tool) on that individual's bike or stereo. Publicize the program; faculty and students will welcome it.

Thefts from residence halls occur most often during student vacations, often perpetrated by service personnel (housekeepers), by workers (contractors or school employees) in dorms, or by other students. Students can lessen that danger by turning in valuables for Security Department safekeeping, by marking personal property, and by using locks.

The security staff can help deter it by patrolling residence halls, by checking individual rooms, and by periodically checking packages being carried from residence halls or off campus.

Vandalism and Graffiti

Damage to property by vandalism or graffiti occurs everywhere these days. It seems a sign of the times.

A 1977 House committee, considering the cost of replacing or repairing school property damaged by crime on campuses, assessed that cost at more than $200 million per year. Because of rising costs, if for no other reason, the cost must be much greater in 1986.

Besides the cost of actually replacing or repairing, however, there are "hidden" costs:

- Cost in manpower, equipment, material, time to clean up the damaged area.
- Increased replacement costs *vs* original purchase costs.
- Extra supervisory costs.
- Extra costs for investigating the incident and for adopting additional security measures to prevent its recurrence.
- Time lost in use of the facility.
- Increased insurance premiums; difficulty getting insurance coverage.
- Cost in adverse publicity to the school (and the effect of bad publicity on the next school bond drive).
- Adverse psychological effects on staff and faculty, on students, on the community.

School property can be damaged by vandalism, burglary, theft, fire, arson, flooding, noxious odors, and by defacing or otherwise damaging surfaces.

Sociologists conclude that those who vandalize a school or harm it by graffiti may include:

- Individuals who have unhappy home lives and find that attacking the school relieves pressures they cannot release at home.
- Individuals who have poor success in studies.
- Individuals who feel out of place in most school activities.
- Individuals who feel a need to "get even" with the principal, with a teacher, with peers, with their families, with the community—with someone.

- Individuals who enjoy violence on TV.
- Individuals involved with alcohol or drugs.
- Individuals who don't like themselves, who have difficulty communicating with others, who have not found an extracurricular activity they really enjoy.
- Individuals who feel they have little to say about what happens at the school and resent it.
- Individuals who feel the school's curriculum or extracurricular activities are boring and have little to offer them.
- Individuals who do not like the community and, as the school represents the community, see an attack on it as an attack upon that community.
- Individuals whose families have not taught them social values and whose teachers also have not had time (or cared) to teach those values either.
- Peer and gang pressure.

Vandalism and graffiti can be controlled by a mix of physical security measures and sociological and psychological programs.

Physical measures might include:

- Fewest possible first floor windows, with shatter-resistant glass or glazing, good locks, alarmed.
- Good exterior and interior lighting.
- Low shrubbery.
- Use of CCTV surveillance, alarms, listening devices, radar alarms to cover hallways and other sensitive areas.
- Security Officer foot and mobile patrols.
- Vehicle patrols by local police.
- Involvement by staff and faculty, particularly during class breaks and in congested hallways and other areas, to maintain order and enforce school regulations. Courage to report perpetrators of crime.
- Use of vandalism and graffiti-resistant surfaces, such as acrylic plastics, in more vulnerable areas (*e.g.*, rest rooms).
- Cleaning of damaged surfaces with special chemicals before graffiti paint or ink dries.
- Community patrols on campus; recruiting nearby residents to keep an eye on the school and report suspicious activities.
- Keeping school facilities occupied and busy with many people during hours the school normally is closed (community affairs, use

by local police department, continuing education programs, community college classes, etc.).

- Establishing mobile home apartments on campus for underprivileged individuals. School offers rent free housing in exchange for occupants' keeping an eye on the school and reporting suspicious activities.

Sociological and psychological programs to reduce school vandalism and graffiti might include:

- Enforced work program requiring perpetrators to clean up or repair damaged areas.
- Holding parents monetarily responsible for their childrens' conduct.
- Punishment for offenders to include:
  - Fines.
  - Suspension or dismissal.
  - Fine and/or work programs.
- Student security patrols exerting peer pressure on potential or actual perpetrators.
- Improvements in curriculum and extracurricular programs to make school more interesting and more challenging to students.
- Job training programs designed for students who clearly will not be going on to college so that the school will have a purpose for them: it will get them a job.
- Keeping students as busy as possible, as much of the time as possible.

─────────

If a school presents an excellent appearance and has a good reputation, it is less likely to be harmed by vandalism or graffiti than a school, perhaps quite nearby, which is not so physically attractive and has a poorer reputation. The school community seem to protect a school they consider "good."

Here is a lesson for school administrators. Run a good school and the school community is far more likely to support you and to protect your school. If vandalism occurs, that community is inclined to police itself and discourage future vandalism.

A clean, neat, attractive campus, manned by a competent and dedicated staff which presents sound, attractive, comprehensive programs and offers a chance for extensive involvement by the local community can be the best "security system" that administrator can find. It also will improve the morale of everyone concerned; save money; allow more school programs; improve the school's esteem before the school community and among its educational peers.

Be assured of one thing: such a school certainly will be a lot more fun to administer.

CHAPTER 11

CAMPUS RETAIL SECURITY

THE OPERATION of campus clothing stores, gift shops, book stores, dining facilities, guest quarters, and theaters, not to mention athletic facilities, may involve valuable material and products and significant school revenue.

The Security Director is concerned that all school property is properly accounted for and that all campus revenue is credited to the school. This places him squarely in the retail security business, on a big campus in a big way.

Volumes have been written about retail security. This chapter intends only to outline the nature and scope of retail security problems on campus and to suggest some solutions for them. Specific solutions will vary from store to store, from staff to staff, but the concepts in this chapter are a start. Most campus retail operations are not large enough to warrant all these suggestions, but many apply for any size operation.

As with other security matters, common sense approaches, applicable to local situations, usually are the best techniques.

Inventory shrinkages (unaccountable losses between assets input and revenue output) result primarily from thefts by patrons but also come from employees' theft, carelessness or incompetency.

Modern retailing emphasizes patrons' "trying on" merchandise (seeing, smelling, hearing, touching, wearing) before buying it. This approach appeals to the patron, but it also make shoplifting more tempting to the professional thief, to the casual pilferer, to the thrill seeker or the neurotic who steals to satisfy a psychological need.

The professional thief is not apt to attack a college retail facility. There are more lucrative targets elsewhere. The average campus shoplifter instead is an amateur, acting on impulse.

The average shoplifter has resisted sociologists' efforts to pin him to a particular group. Juveniles predominate among those apprehended, and males exceed females. Beyond that, no clear pattern emerges. There is no predominance among racial or ethnic groups and, if many shoplifters are from poorer groups, many also are from well-educated, religious families. The shoplifter steals because he can't pay for the items, because the taking reduces stress or anxiety, because he relishes the excitement the risk provides.

His methods vary as much as his identity and motive but may include:

- Switching price tags.
- Grabbing merchandise and running.
- Palming small objects or concealing them in books or newpapers.
- Using umbrellas, purses, shopping bags, brief cases to conceal stolen items.
- Using packaged items reopened and repackaged to conceal additional, unpurchased items.
- Wearing stolen items from the store (openly or under loose fitting clothing).
- Using slit pockets in coats to attach articles to hooks or extra large pockets inside the coat.
- Using false store packages with openings through which he conceals stolen items.

The best defense against shoplifters is a systems approach combining use of the security and sales staffs, electronic surveillance and alarm systems, physical and psychological barriers between the patron and temptation and education of the staff and public to punishment of apprehended shoplifters.

Physical barriers might include the arrangement of display cases, clothing racks, doors, and cash registers to make sensitive items or cash difficult to steal. The longer it takes, and the more the potential shoplifter is exposed to observation by the store's staff or the public, the less likely he is to risk being caught.

A combination of physical and psychological barriers might include setting cash registers behind counters, jewelry displays in locked cabinets, stock rooms apart from sales areas. Symbolically, as well as physically, they separate these items from the patron's normal grasp.

Garments may be protected by electronic sensors which use interrupted radio waves to magnetic fields to trigger an alarm when that garment is moved from its authorized area. Their presence deters potential shoplifters, and deterrence is better than apprehension.

Surveillance of sales areas is critical. It may be by the security staff (closed circuit television, viewing stations, plain clothes detectives) or by the sales staff.

Sales personnel should wear distinctive clothing (uniforms, badges, etc.) which identifies them as store employees and make no attempt to hide their concern for their areas of responsibility. If their presence and interest deter the shoplifter, apprehension and the difficulties of prosecution will never be necessary.

Surveillance is helped by assuring that aisles are straight, wide, uncluttered; and by rearranging racks or mirrors to prevent blind spots.

Surveillance by the security staff may be by closed circuit television (CCTV). The CCTV system should be clearly visible to the shopper and convince him that he is being observed (red flashing light on cameras, tracking cameras, TV screens open to the public view and showing customers). Dummy cameras, mixed with operating ones, add to the conception of an extensive surveillance system.

To be effective, operating cameras must be constantly monitored by a Security Officer while a second Security Officer is available to respond to a suspicious activity.

Other limitations on CCTV surveillance include:

- Limited capability of the camera (lighting problems, size of area covered, vulnerability to electronic disturbances, lack of photographic detail).
- Problems in camera placement, blind spots, vulnerability to lens damage).
- Camera's limited mobility (as displays change surveillance requirements change; different camera mountings and outlets, however, may not be available).
- The need for video tape recorders to document thefts.
- The system requires frequent servicing.

CCTV surveillance is not foolproof. It is a deterrent, however, and every bit helps.

Public education to discourage shoplifting might include posters and signs, such as:

- "Stop Shoplifting; Everyone Pays for It."
- "Shoplifting is a Crime."
- "Shoplifters Will be Prosecuted! Don't risk a Criminal Record."

- "We Reserve the Right to Inspect all Packages Carried from the Store."

Posters and signs should emphasize that shoplifting is a crime, is degrading, causes everyone to pay higher prices, and that apprehension will be followed by public prosecution.

The message also can be carried in campus publications, in pamphlets, in spot ads on the school's television station, in speeches and rap sessions between school administrators and students.

Local police can informally survey campus retail facilities and offer security suggestions or participate in student orientations (talks, puppeteering, films, role-playing, etc.) on good citizenship. To local police, preventing shoplifting is a priority project (more so than the prosecution of an apprehended shoplifter). They'll help all they can.

The Security Director's best help, however, lies with the retail security and sales staffs. They must be trained in retail security measures:

- Scope of shoplifting problem.
- Security operations; summoning security help.
- Employee sales operations (cash registers, credit transactions, inventories, stockage, tagging, etc.).
- Inventory controls (receiving, shipping, stocking, displays).
- Legal ramifications (shoplifting, apprehension, arrest).
- Store policy regarding shoplifting.
- Anti-shoplifting procedures (see below).
- Recognizing potential shoplifters (see below).

The sales staff can help prevent or deter shoplifting by:

- Keeping their departments neat, clean, uncluttered; avoiding empty hangers and disarranged stock.
- Arranging counter displays in patterns so incomplete sets will be easily noticed.
- Arranging displays to lessen temptation.
- Keeping display cases closed or locked.
- Using mirrors to view blind spots.
- Fitting rooms: clear of merchandise; single aisle entrance; walls and curtains of changing areas exposed one foot above the floor; allowing a customer to take not more than two (2) items to the fitting room at a time.
- Displaying only one item of a pair.
- Fastening plugs on electrical fixtures to make concealment more difficult.

- When possible, keeping articles (*e.g.*, luggage, appliances) fastened to racks.
- Circulating as much as possible; not congregating with other clerks or customers for excessive periods; checking fitting rooms carefully.
- Being alert for nested, undeclared, or secluded items.
- Sales supervisors' observing customers closely; checking price tags, arrangement of displays, handling of cash and credit transactions, fitting rooms.
- Learning the habits of frequent customers; identifying those who require special attention.
- Greeting all customers but serving only one at a time.

Suspicious appearance and behavior by customers include:

- Appearance
  - Loose clothing; coat over shoulders or arms.
  - Unseasonable clothing.
  - Carrying empty or open bags; obviously used packages; knitting bags; briefcases, unusually large purses, newpapers, magazines, or books.
- Behavior
  - Agitation or extreme nervousness.
  - Constant walking up and down aisles.
  - Walking around with merchandise in hands.
  - Covering articles, folding them, dropping them to the floor then returning them to the counter, concealing them.
  - Removing or switching price tags.
  - Frequent glancing about.
  - Continuously refusing sales help.
  - Trying on jewelry or clothing and leaving it on.
  - Entering the fitting room with packages, with various sizes of clothing, with obviously wrong sizes, with several of the same items.
  - Signalling or otherwise working in concert with another individual although the two are not together.
  - Frequent trips to the store without making purchases.

Protection against shoplifters is made tougher because their apprehesion is fraught with legal land mines. Effect an unwarranted apprehension or arrest, and you can virtually count on being sued.

To be legally safe, the apprehender should consider the following criteria:

- He has reasonable ground to suspect that the individual has committed or is committing shoplifting.
- The apprehension was conducted in a reasonable manner.
- After apprehending the suspect, the apprehender behaved in a reasonable manner: privacy during questioning, searching, questioning technique, statements made by suspect, delivering suspect to the police.

The key word, of course, is *reasonable*. For example, he reasonably believes that the perpetrator means to steal because:

- He has personally seen the shoplifting done.
- He has another witness to the shoplifting.
- More than one article has been taken.
- The article(s) is concealed (carried in a place or manner in which such merchandise normally is not carried), and he knows where it is concealed.
- He has seen price tags switched or removed.
- He has overheard suspect's conversation to the effect that shoplifting is intended.
- His search has recovered at least one of the stolen items.

Apprehensions should be effected outside the store: there is stronger proof of intent and, if there is a confrontation, it's best done outside.

In making an apprehension, the Security Officer should:

- Use as much courtesy, firm gentleness, persuasiveness, other aspects of professional behavior as possible (to avoid confrontations and to show a reasonable approach).
- Should not touch the suspect until he has spoken to get his attention. Then he should quietly direct the individual back to the store, taking his arm in a firm but gentle manner, if necessary, to suggest it.
- Talk quietly with the suspect as they return to the store, eliciting information and lessening anger or tension that might cause a confrontation. Information of a personal nature (first name, family situation, work, etc.) can be a lever in eliciting a confession.
- If the suspect becomes hostile or denies taking the goods, do not release him but get him back to the store.
- In the security area of the store, question and search. If possible, have a witness present. When interviewing a woman, *always* have a witness present. If a search is involved, and the suspect is a woman, have another woman conduct the search.

- Have the individual remove any coat and empty pockets or any containers he is carrying. Don't touch money but inventory the contents. Be alert for locker keys, motel keys, etc. for the location of possible other stashed goods.
- Get a verbal confession to his taking at least one item; then have the suspect prepare a written statement, reread it, sign it; witness signs also.
- Follow store policy regarding prosecution. If appropriate, notify local police to make the arrest.
- As soon as possible, prepare written reports and submit with other evidence to the security supervisor.

Store policy regarding shoplifting penalties should be written, publicly displayed (if prosecution involved), and enforced.

Security or sales staff members who apprehend or assist in the apprehension and confession or prosecution of a shoplifter, should be rewarded with incentive bonuses, time off, or public recognition.

Employee Theft

On the average, nearly one-third a retail store's inventory shrinkage may come from theft by its employees.

Studies indicate that employee thefts occur most significantly among:

- Young male Caucasians.
- Never married.
- Living in higher income households but contributing little to that household's total income.
- Occupying jobs providing easy access to merchandise or cash.
- Employees dissatisfied with their supervisors, with their jobs, with their company.
- Employees occupying lower paying, lower status jobs.
- Employees who believed they'd never be caught and, if caught, would never be prosecuted.
- Employees of firms which do not have aggressive anti-theft programs or policies.

Likely candidates for employee thefts might be:

- Those living beyond their means (housing, car, clothing, etc.).
- Compulsive gamblers.
- Those involved with drugs or with drinking problems.

- Financially irresponsible individuals.
- Chronic complainers who feel rejected or unappreciated.

Indicators of employee thefts include:

- Large, unexplained inventory shortages, particularly among more sensitive and more expensive items (calculators, jewelry, etc.).
- Decline or unusually small increases in cash or credit sales.
- Profit declines.
- Increase in expenses.
- Increase in the return of sales.
- Cash register audits revealing:
  - Gaps in transaction numbers.
  - Unusually large number of overrings, voids, or no sale transactions.
  - Frequent and clear pattern of overages or shortages of merchandise or receipts.

Theft by retail clerks might be of merchandise or cash. It can be done by:

- Overcharging a customer.
- Ringing up a "no sale" or voiding a sales slip, then pocketing the money.
- Failing to ring up a sale and no receipt issued to the customer; pocketing the money.
- Price cuts to a confederate.
- Passing merchandise to a confederate or carrying it away himself.
- Swapping stolen merchandise with another clerk.
- Stealing cash from a common drawer cash register.
- Stealing from delivery trucks, warehouses, stockrooms.
- False inventory reports, invoices, or orders.
- Cashing known bad checks for a confederate.
- Shoplifting from other departments.
- Wearing jewelry or clothing as a "model," but not returning it to stock.

Measures to discourage theft by sales employees might include the following:

- Pre-employment background investigations emphasizing character and work habits.
- Investigation of work and living habits of suspected employees.
- Periodic polygraph examinations, when permitted by law.

- Better supervision of all aspects of employee operations.
- Management spot checks: stock inventories, sales books, receiving areas, shipping operations, credit transactions, customer billing, etc.
- CCTV observation of work areas (hindered by employees' knowing cameras' vulnerabilities (blind spots, lighting, etc.).
- Undercover operations.
- Honesty shopping (Security Officer "patron" makes purchases to test sales procedures).
- "Salted" merchandise (Security Officer "patron" returns "found" item and observes whether it is properly credited, returned to stock, etc).
- Prosecution or instant dismissal of employees apprehended stealing.

Shipping and Receiving

Shipping and receiving operations also result in losses. They occur when employees are careless or misled in properly checking goods shipped or received or when they deliberately steal merchandise.

Shipping and receiving operations should be kept separated from stocks and sales areas and from each other. If they cannot be done in separate rooms, at least the general common area should be divided to keep shipping and receiving operations separate and distinct.

The receiving area should be well-lit (outside and inside). Exterior doors should be sturdy and secured with locks and alarms, with closed circuit television surveillance. Delivery personnel should be restricted to the immediate "dock" area and not be allowed to move freely and unsupervised through shipping, receiving, and stock areas.

If possible, goods should never be received through front doors leading to sales areas. Delivery through front doors leads to confusion and increases the danger of theft, may cause injury to clerks or patrons, and presents a poor appearance. Use rear delivery areas.

The receiving area should be separate from the store's trash collection area. If they are adjacent, there is danger of shipping cartons being mistaken for trash or intentionally hidden with trash for later retrieval.

Unless goods move immediately from the receiving area to the sales area, it may be necessary to construct sturdy mesh-protected and locked cages for protection of smaller, more valuable items (cameras, jewelry, calculators, etc.).

If possible, receiving should be scheduled only during specified hours to reduce confusion and to allow the best employees to handle it.

Receiving should be handled by or under the supervision of a very competent and loyal employee who fully understands that everything he says is in a shipment is going to be entered in the store's inventory and be charged to the store.

He must be thoroughly familiar with all forms used for receiving material. Purchase orders, *not packing slips,* must be used when verifying shipments.

If possible, two employees should count received cartons separately. If that is not possible, let the deliverer do an independent count, the receiving employee another. If the two do not agree, the receiver should stop, complete a discrepancy report, and have it verified by his supervisor.

To receive the shipment, the receiving clerk must verify the total package or unit count in the "dock" area. Later, inside the receiving area, he must open each carton and verify its contents (totals, type merchandise ordered, etc.). If there is a problem, he again must prepare a discrepancy report and notify his supervisor.

In counting cartons piled in blocks, he should be wary of cartons missing from the core of the block and should limit his count to not more than 25 blocks; a larger count is apt to become confusing.

Purchasing agents (store buyers, etc.) should not be involved in receiving procedures.

The receiving clerk must not allow himself to be distracted by drivers' conversation or questions. He must never take another's word for the count but must verify it for himself.

He must not leave material unsecured on the loading dock. If it is his break time, he should complete the count, move the merchandise inside, and secure the outer door before taking his break.

He must be alert to tricks drivers may attempt, such as:

- Fast counts, fast talk, questions, or idle conversation to distract his count.
- Signs of impatience or anxiety by the driver to get the count completed so he can be on his way. Make him wait.
- Closely packed clothing on racks (several hangers may be empty).
- Items left behind in the truck during the receiving count. If the shortage is detected, the driver returns to the truck and "finds" the item he'd "overlooked." This is where portable spotlights to illuminate truck interiors help.

- Piling small cartons in stacks with hollow centers.
- Using cartons to support one end of an unloading conveyor, then loading those cartons back in the truck after the count is completed.

Deliverers may confuse or fool receivers. It can be an honest mistake; drivers can be careless too. But, if they have a will to steal, they'll find ways to do it and have no regard for the receiving clerk whose signature attests that he received the missing items.

Just as receiving functions should be separated, so should shipping procedures. Stockrooms and shipping areas should be apart. The person ordering the shipment should not take the merchandise from the stock room or prepare it for mailing. Order fillers should not be given the customer's name and address. Shipping clerks should pack and wrap merchandise before they receive the customer's identity.

Periodically, supervisors should open and spot-check random packages and stock areas. They should review mail pickup procedures and randomly audit purchase order addresses. They might also initiate error control programs with cash incentives or other rewards for error-free work.

In all campus revenue producing activities the Security Director should encourage use of an unbroken chain of receipts whenever merchandise or cash transfers from one person's responsibility to another's.

Cash should not be left unsecured and unattended at any time. Small safes should be secured to permanent fixtures and the area should be well-lit. When possible, all cash containers should be emptied at night with revenues deposited in the local bank by Security Officer-escorted messengers. Empty cash registers or other cash containers should be left open with a sign stating that the container has been emptied.

Steps to control inventory shortages include conducting a survey to identify problems; developing security consciousness and cooperative support in an anti-theft program among employees; improving merchandise flow, merchandise controls, merchandise layouts and stockage; being constantly alert to sensitive or new problem areas.

Retail losses can be cut. Doing it requires work, imagination, and the willing involvement and support of both the retail security and sales staffs.

CHAPTER 12

LIBRARIES, MUSEUMS AND GALLERIES

SECURITY problems that beset campus libraries, museums, and galleries include the following:

- Vandalism of property (broken fixtures, windows; stopped drains; fumes; graffiti; etc.).
- Vandalism of collections (collections torn, defaced, etc.).
- Theft (of collections and property).
- Fire or water damage.
- Damage from vermin or rodents.

We will focus on problems related to libraries, but problems and solutions often apply also to galleries and museums.

Despite the use of many accepted security techniques, libraries, museums, and galleries remain tempting and vulnerable targets. This is because of the nature of their collections, because they are open most of the time, because they offer freedom of access, and because their staffs are limited in their ability to protect the large complexes these facilities represent. Collections are particularly vulnerable. Often they are priceless in value and a single mark or tear or stain may ruin their artistic value. And an individual intent upon defacing or stealing usually is very difficult to detect.

One need only spend a fruitless hour searching a library's book stacks for books not checked out yet missing to realize the magnitude of the problem.

An individual has a research paper due and finds it more convenient to conceal the book than to check it out. Or he may deface pages with underlining or cryptic notations or simply tear them out.

Losses like these are staggering to a library. Most libraries have no idea of their extent. Nor can they pinpoint exactly how they occur or by

127

whom. One thing is certain, losses occur in every library, and every patron suffers because of them.

Part of the problem lies with limited funding for security. The library cannot afford alarms systems, CCTV, extra security guards. The staff is small and must cover a large facility. They are busy most of the time. And they often do not believe that security is their responsibility. To many of them, security measures are the tail wagging the dog. Librarians are there to spread knowledge, let someone else worry about the collections. So they say. Besides, the staff must deal with library patrons. And patrons can become very annoyed with delays in processing books and may resent an inspection of their briefcases, purses, or other containers. They'll declare the search to be an abridgement of their personal right to privacy and an insult to their honesty and integrity. If found with books not properly check out, their defense is, "Well, I forgot all about those;" or "I intended to have them checked out when you interrupted me." Caught marking a text or tearing its pages, they're apt to shrug, "Everyone else does it; why shouldn't I?" Worst of all, the library administrator may refuse to punish even the worst offender. Realizing that, one can understand the library staff's avoidance of confrontations. Still, despite all the security gadgetry available, the staff remains our best defense and, like it or not, must be involved. It's part of their job.

Security guards, hired to patrol near the library's exit, are far from foolproof. They are apt to be of limited ability, working for limited pay. Their motivation may be far less than that of the average library staff member. After all, books are not personal things to them. And books are so easy to hide: coats, skirts, briefcases, purses, knitting bags, folded newspapers, all offer good concealment. Besides, there are other exits which are not guarded or windows easily opened above shrubbery which offers a convenient place to conceal a book for later retrieval. And security guards don't relish confrontations with irate patrons any more than the average library staff member.

Guards, like other security measures, help but they are not infallible and, on their best day, will have limited success at stopping theft of books.

Other elements of the security system (alarms, CCTV, sensitized book inserts to trigger alarms, etc.) may be poorly designed or may not be allowed to operate as they should. The library can buy an expensive CCTV surveillance system, for example, but if the camera's lens is masked by desks, books, or other library paraphernalia, or if there is no one to properly monitor that camera, the system is defeated.

Most likely, the school library will operate under severe budgets that preclude sophisticated security equipment. Its director must rely almost entirely on his staff to handle the security problems that are sure to be there.

Unfortunately, from his point of view, the staff will be very busy with activities in a large complex of different rooms, each room itself broken up by books, desks, chairs, and other library furnishings. Direct observation is difficult, and they don't have much time to concentrate on security matters anyway.

The library staffs' busyness also may cause individuals to be careless in their own procedures (registration, in/out checking, shelving), so they lose track of books still within the library.

Any school library offers more than enough problems to go around. As with all security matters, there is no absolute solution. Nothing is foolproof. But damage or theft of library property can be discouraged and contained by the use of certain security techniques, most of which are not sophisticated or costly at all.

School libraries, once housed in separate buildings centered on the campus, now more often may be dispersed among a half dozen academic departments or other activities scattered throughout the campus. Even the name has changed. Now they may be combined with classrooms and termed "learning resource centers," or "instructional material resource centers." In any event, there is a tendency to mix libraries with other activities in a particular building so they may be close to particular academic disciplines.

The security system for a particular library must be tailored to that library's configuration and operational procedures and, if it shares building space with other activities, must be tied into the building's overall security system.

The Security Director must use a system approach, combining human and mechanical skills available in the best possible way for that library. The system has to be cost effective and must not interfere with the academic operation one bit more than absolutely necessary. Any sign of "the tail wagging the dog," and the Security Director will have problems with the library staff. His plan must recognize that the library staff can be only a part of his defenses. And they must take in consideration the operations of other staffs active in that building.

His first line of defense, as with most security systems, is the library's exterior walls. He might keep in mind the following suggestions:

- Reduce the number of entrances and exits to the bare minimum necessary.
- Assure that exterior doors are sturdy, with good locks and hinges that cannot be removed from the outside.
- All openings in the facility, including any on its roof, must be locked (when not in use) and alarmed.
- Ladders or other means of access to the roof should be secured.
- First floor windows should be held to a minimum. They should be difficult to reach from the ground and, if open, should be protected by mesh or screen. Impact-resistant glass or glazing should be used on all windows susceptible to damage.
- There should be reasonably high light levels at all entrances and exits. All exterior lighting should be not less than 2 foot candles' intensity at ground level. Light fixtures should be set for automatic timing or hand control and should be screened to resist damage by hurled objects.
- Exterior electric or telephone wires should be in underground conduits to deter sabotage or other damage.
- Any openings in the building's skin (air ducts, conduits, water pipes, book return windows, etc.) should be provided with protection against their use to spread fumes or fire inside the building.
- Shrubbery or trees should be of a thorny type to discourage their use for concealment. Shrubs should be no higher than 2' above ground level, and trees should not be planted near the building.
- Library entrances should be kept to the absolute minimum necessary (perhaps an entrance/exit combined with a freight/staff entrance/exit combined); should be marked, equipped with locks and panic bar releases, alarmed, under observation when not locked. Entrances should provide for use by handicapped individuals. Doorways should use break-resistant glass or glazing material.
- All fire exits should be alarmed and marked for emergency use only.

Service counters usually are divided into two general activity areas: those processing in-bound patrons (book check-in), those processing out-bound patrons (registration, fines, check-out).

Their location and configuration, important security factors, may depend upon such considerations as:

- The library's configuration.
- Staff availability.

- Size of library and its staff.
- Patron traffic.
- Routes necessary for any book detection system.

The staff should have as much visual contact with patrons at the entrance/exit as possible. This is enhanced by avoiding confusion in patron lines. If possible, keep patron lines separated. It reduces confusion (also lessens the temptation to take a book without checking it out) and speed processing. This may require two or more service desks (in; out; registration; fees; information) or clear delineation of areas of service on the one large desk available. Signs should reduce congestion and waiting at any one service area.

However service desks are arranged, the basic idea is to prevent congestion near exit areas, thus deterring an individual's slipping out, undetected, with library property. Service areas also can be improved by restricting the space between the service desk and the opposite wall (or a floor furnishing) to channel and slow pedestrian traffic.

A Security Officer or library staff member should be near the entrance/exit area to oversee operations and to check packages, answer questions, etc.

In enforcing library procedures, particularly when a staff member confronts an individual suspected of misdoings, the following suggestions apply:

- Staff members should never rely on another patron's word but should see for themselves.
- Staff members should conduct a reasonable investigation before detaining an individual.
- Only more senior staff members should have authority to detain an individual.
- Individuals should never be detained without reasonable cause. Legal ramifications are enormous.
- Staff members should never get involved in a shouting match with patrons; let the supervisor handle those individuals.

Public service areas (book stacks; reference rooms; areas containing microfilm readers, typewriters, copiers; rest rooms, etc.) are real security risk areas and must be treated that way.

Hazards in the book stack areas, for example, include their use as a lovers' lane, their use by "peeping Toms," the site for rape and robbery. Risks in other public service areas include damage to or theft from machines, damage to library property or, again, harm to patrons.

The answer lies in such basics as requiring the staff to move through public service areas as often as possible, eliminating clutter and improperly placed furnishings which impair security observation, and, if available, use of CCTV and mirrors.

Patrons in reading rooms should be in clear view. It helps if the reference librarian's desk can be elevated to give a more unobstructed view of the reading area.

Precautions for use of manuscript repositories might include the following:

- Carefully identify all users and maintain a users' log.
- No briefcases.
- Patrons permitted to work on only one box or one unit of manuscripts at a time.
- Encourage folders and volumes being left flat on tables, not leaning upright.
- If a patron leaves the room temporarily, any material he takes out should be checked; papers should be placed back in folders during his absence.
- Returned folders are immediately checked for contents.
- Manuscripts which can be marked for identification without harming their value are marked by embossing, punching, perforating, or stamping.

Rare manuscript collections are particularly vulnerable. The least tear or mark may destroy much of their value. The risk of theft is increased by the library's not being able to mark the items as library property. Procedures that will help include:

- Restricting use to individuals who have established a clear need to use the *original* documents and who have been properly identified and logged into the rare manuscript area.
- Using xerox or microfilm copies for research purposes. If it is the information on a manuscript that is critical, not the manuscript itself, use a copy.
- Surveillance by a staff member of use area.

Loss, destruction, or significant damage to card catalogues also can be very harmful to a library. Loss or damage can be caused by fire, theft, tearing, or attack by paint, glue, acid, ink, or similar substances. Somewhere there should be a duplicate or microfilm copy of each card so that loss of the original card will not put the library out of business. Card catalogue containers can be secured at night by locks (*e.g.,* locking bar along length of card catalogue cabinets). At other times they should be supervised by a staff member. For these reasons, it is best to place card catalogue cabinets near a staff desk.

Public service equipment (typewriters, copy machines, microfilm readers, tape recorders, projectors, etc.) should be made available in well-lit, open areas under staff supervision. Coin operated equipment should be emptied at night with coin boxes left open and a sign drawing attention to that procedure.

Staff work areas should be protected by locks (perhaps with card-key combinations for the staff) and should have "Staff Only" or "Off Limits" signs, CCTV surveillance, and, when appropriate, panic alarms to summon help.

Depositories for return of books when the library is closed, if built into the library's exterior wall, are very vulnerable to vandalism or other damage and offer a tempting opening for someone to deposit animals, trash, or other unwanted material or to attack the library by fire or the spread of smoke, odors, noxious fumes. Normally it is better to use larger, heavy gauge metal containers (similar to those used by the postal services) at curb side, rather than have a book depository built into the building itself. If the depository is to be a part of the building, however, it should be isolated by fire-proofed walls and equipped with smoke and heat detectors. There also should be a floor drain nearby to handle any attack by water from outside the building. If all this costs too much, use the curb side stand.

Whether an exterior container or a building depository is used, it should be locked during hours the library itself is open.

The library should be equipped with a safe (or vault for large libraries) to protect rare collections, film, video or audio tape, and money from vending machines. The safe should be in a well-lit area, secured by chains or bolted to the floor, to a radiator, or to some other permanent fixture. It should be alarmed. Alarms should include heat and smoke detectors. A vault should be well-lit, equipped with intrusion, smoke and heat detectors, have separate temperature and humidity controls. It also should be equipped with an emergency power source and with emergency communication.

Custodial areas should be thoroughly checked by the staff, particularly before closing the library, as they offer excellent places to conceal an incendiary or explosive device, stolen items for later retrieval, or an individual intent upon hiding until the library is closed.

Custodial areas must be kept free of debris, paint, or other flammable material; must be neat and uncluttered; and should contain a floor drain to contain any water damage.

Fire represents another significant hazard to libraries. Library collections are slow to ignite but, once ignited, are difficult to put out. During a fire, book stacks can act as natural flues, making a fire much worse. Fire preventive measures suggested elsewhere in this text apply well to libraries, museums, and galleries, but these facilities also offer problems peculiar to themselves.

Libraries located near other buildings of somewhat volatile nature (industrial plants, certain research facilities, etc.) should have a fire resistant roof.

Interiors should be designed to confine any fires; fire walls, fire doors, HVAC systems can minimize its spread.

Records storage should be parallel to aisles, rather than perpendicular to aisles, as this will help confine the fire.

Representatives of local fire departments should be given complete tours of the facility to insure that they know its physical configuration, control systems, and location of critical areas, controls and equipment.

The library must be kept neat, clean, uncluttered, free for quick access and exit. Desks, closets, waste baskets, and trash cans must be managed properly. Wooden trash cans should be avoided, and all trash cans should be emptied often to avoid overflow.

Separate areas should be established and enforced for smokers.

Libraries should be equipped with heat and smoke detectors and halon, CO_2, or dry chemical extinguishers. As most library staff members are women, smaller extinguishers are appropriate. Extinguishers must be inspected monthly to verify that they are where they should be and are serviceable. Prominently displayed, legible signs should indicate fire alarms, exits, extinguishers, and use of fire fighting equipment.

Staff fire teams should be organized, equipped, and trained. Fire plans should be written, taught, rehearsed, and enforced. They must be current. An obsolete plan on a piece of faded paper will be of little value, for example, if the physical configuration (rooms, exits, stairwells, extinguisher locations) have changed.

Training for in-house fire fighting teams should include:

- A thorough knowledge of the library's fire prevention and fire fighting plans.
- An orientation on individuals and duties assigned to various teams on each shift.
- The importance of the collections; most sensitive areas.
- The proper selection, location, and use of fire extinguishers (to include actual practice).

- Routes to exits; fire doors and use of fire doors.
- The proper carrying and use of any other fire fighting equipment.
- The location of standpipes, hose stations, light switches, other electrical controls (temperature, humidity, lighting, alarms).
- The use of halon or other extinguishing systems.
- Alarm systems (type, how to operate, activating, locating a triggered alarm, resetting triggered alarms).
- Availability and use of emergency communications equipment.

Members of the staff fire fighting team should be rewarded for participating in training sessions, drills, emergencies. If you can't pay some stipend, offer a free lunch, compensatory time, public recognition.

Monthly fire inspections should be conducted to detect and correct discrepancies, such as electrical circuit overloads; frayed or otherwise damaged electrical cords; presence of flammable materials; or blocked standpipes, sprinklers, fire exits.

Where plumbing exists in the library (kitchen, rest rooms, custodial areas, etc.), there should be floor drains to prevent or contain flooding.

The most vulnerable items in the library are the books themselves. Besides being stolen, they can be damaged by deliberate defacing, spilled ink, paint, or similar substances; pages folded, torn, glued together. Such vandalism is easy to do and difficult to detect.

The best defense remains an alert library staff. Certainly the staff has the best chance to prevent theft of books.

Patrons steal books because they feel they need them for their work or for research purposes or simply because the risk involved represents a thrill to them.

Besides an alert library staff's frequently moving through public service areas and surveillance by CCTV, the best defense against book thefts is the library's check-out procedure. The configuration of the check-out desk and adjacent wall should restrict patrons to one at a time. The books most likely to be stolen can be treated with sensitive material, available through a library supply distributor. Imbedded in the book, it activates a warning system in the check-out area if it has not been cleared by the check-out procedure. Such a system requires the patron to pass between the desk and a screen (part of the facing wall or of another piece of furniture). The system works adequately but is expensive and should be reserved for the most sensitive books.

Another real danger to library collections is damage from vermin or rodents. Moths, beetles, rats, and mice like dark bookshelving and make nests of newspaper or similar material common to a library. They are most likely to be found among collections seldom used (often the more valuable collections).

Maintenance personnel must close unwanted openings around pipes, floors, or walls. Vermin don't like light; windows or lights discourage their presence. Shelving should have open bases and be four or more inches above the floor and free of standing walls. If mold is a problem, dehumidification equipment should be installed. All personnel must be alert to the first sign of vermin or rodents. Caught early, appropriate pesticides or other control material can quickly eliminate the problem.

Museums and Galleries

Museums and galleries require protection similar to that of libraries. One might add, however, that museum and gallery collections usually are more valuable and more difficult (or impossible) to repair or replace than a library's holdings.

There also is great concern among museums and galleries for the institution's reputation. If a valuable painting or sculpture is stolen or damaged, and it appears that the institution's staff and security systems are not all they should have been, that museum or gallery will find its insurance coverage more costly and probably more difficult to get. It also will find it much more difficult to borrow another museum's collections for its own showing.

Threats accentuated in museums and galleries include vandalism, terrorism, robbery, and internal or external theft of collections.

The intensified problems are reflected in increased insurance costs. Other factors affecting those costs include:

- Quality of the guard service.
- Quality of the alarms systems.
- Quality of the museum or gallery staff.
- Registration procedures.
- Fine arts loan procedures.
- Quality of packing and shipping, receiving departments.
- Error control procedures.
- Fire prevention systems.

A good Security Director can influence every one of these factors.

Basic library security techniques, as suggested earlier, apply to museums and galleries. For the remainder of this chapter we will concentrate on only one measure, albeit a key one, common to all three facilities: the security guard.

Security Officers must be neat, clean, pleasing in personal appearance. They must convey a professional demeanor. This does not mean that they cannot smile and greet visitors: it does mean that they should look and act the part of a professional law enforcement officer at all times.

Much of the Security Officer's time will be spent answering routine questions about the facility itself, about the collections in general, about a specific item. They must be courteous at all times. If they cannot answer a question, they should simply say so and direct the questioner to someone who can.

For the museum or gallery guard, the first patrol of every shift is critical. He should check every gallery and every corner within every gallery. He should examine each piece of art to be sure that all items are present and undamaged. He should be alert for any items which have been removed and, where no art removal notice has been passed to him, immediately report it.

Throughout his shift he should move constantly from gallery to gallery, avoiding being distracted by visitors, staff members, or other guards. Except during authorized breaks, he should not congregate with other guards.

During every tour of the facility, he should carefully check closets, work areas, and storage areas.

He should prevent visitors from touching the collection. This applies to statues as well as to paintings. If he observes a visitor acting suspiciously, he should subtly make it clear to the individual that he is being observed. Always, deterrence is better than apprehension after the damage has been done.

Adults or children who are unruly or otherwise violate gallery requirements should be politely corrected. If that does not work, they should be asked to leave.

Visitors carrying parcels, umbrellas, briefcases, or other containers not permitted by gallery rules, should be challenged.

Persons with cameras should be asked for their camera permit.

Lost and found items should be turned over to a supervisor and noted on the Security Officer's shift report.

Guards should always be alert for safety and security hazards: unlocked doors or display cases; broken or malfunctioning equipment; improper electrical appliances; flammable material; damaged, missing, or improperly displayed art items.

Theft of an art object might be indicated by:

- Observation of theft taking place.
- Smashed glass or cabinet in a display area.
- Painting cut from its frame.
- Art item missing from its customary place.

If the Security Officer suspects an art theft, he should immediately:

- Notify his supervisor.
- Inform other guards by phone.
- Secure the exit on the floor where the item is missing.
- Screen all persons in the area.
- Screen persons leaving the facility.
- Secure the possible crime area.

No guard should handle or move art objects except when that object is in immediate danger or when he has been ordered by the curator or director of the gallery to move them.

Guards must use the minimum force necessary to insure compliance with museum rules. They must understand that it is not necessary to arrest a rule breaker. As a matter of fact, a false arrest almost inevitably means a law suit. Ask the individual to leave. If he refuses, report it to the nearest supervisor and let him handle it.

A guard on night patrol who encounters a burglar, however, or a gallery guard who encounters a vandal or a thief in the act, must take immediate steps to stop the illegal action. Again, using the minimum force necessary.

Guard reports are very important to any secured facility but are especially critical in a museum or gallery where the institution's reputation is so important. Guards must be taught to report every unusual incident: clearly, legibly, completely, and accurately before they end that shift's tour of duty.

Libraries, museums, and galleries pose substantial problems for the Security Director. Many of the problems are unique to that type institution, but common sense application of basic security measures and the willing involvement of staff members in enforcing the security plan can go far to reduce security and public safety hazards.

CHAPTER 13

CROWD CONTROL

A MAJOR PROBLEM the campus Security Director must face and master is control of crowds attending campus special events: dances, rock concerts, student mixers, graduation exercises, academic convocations, athletic events.

His responsibility is to assure that traffic to and from the event moves as smoothly and as quickly as safety permits, that vehicles may be readily parked with reasonable assurance of their safety and security, and that participants may feel that they are in a safe environment where they can relax and enjoy the activity.

The Security Director won't be responsible for every facet of the event but, as his responsibilities cross lines with all departments and all activities, he and his staff are going to be involved (like it or not) with many activities they well may question as their department's responsibility. It just happens that way. Some of these activities will become evident in this chapter.

Good control of crowds attending campus events doesn't just happen. It involves three distinct operations:

- Development of control plans.
- Enforcement of those plan during the event.
- After action analysis and modification of the basic plan to correct weaknesses/errors and improve the plan for the next event.

Planning is helped by as much advanced warning as possible. Well in advance of an event, school administrators should alert the campus Security Director. As the event nears, they should add to the detailed information available to him about it: type event, date, time, location, expected crowd, type crowd (students/outsiders), schedule of events, anticipated problems, etc.

Administrators are supposed to routinely provide this detail. It's called coordination. Sometimes it doesn't work. Administrators feel they are too busy to be bothered with security matters, or they fail to see the relevancy of some items to security, or they simply forget.

The Security Director must continually sell himself, his department, and their services to the school. He must earn the respect and recognition of his fellow department heads. He must be readily available when they seek him and take time to seek them out for informal conversations that establish and maintain close personal rapport. Once he has won their confidence, they instinctively will keep him informed of matters affecting their activities. Not because they've been told to keep him informed, but because he's taught them it's to their advantage to do so. He can protect them; he can make their departments look good.

From his first warning of the upcoming event until his after action analysis is completed, he should keep written records of all key matters concerning that event and his department's role in it.

He should develop Standing Operating Procedures to use for each type campus activity, with essentially minor alterations for changed circumstances. Often he will be able to adapt a plan which successfully protected one campus event to another of a similar nature.

Early in his planning phase he should have a clear understanding of funding for the event. To whom will his extra Security Officers and extra equipment, overtime, etc., be charged? If to another department, is there a limit on his spending authority? If to his department, have adequate additional funds been provided and, again, are there special limits to his authority to approve expenditures?

The campus Security Director must have the final word on security arrangements. He knows the field; it's his responsibility. Either he is given adequate resources and authority to provide a safe environment for the school event or he must look to his supervisor for help in resolving the problem. If he feels he cannot provide adequate security and public safety, the event should be relocated to a safer location, be postponed, or be cancelled altogether. In any event, he must have the final word on what is "safe," and his decision must be backed by the school's head.

That is seldom an envious position for him. Often it is a judgment call based solely on his experience and instinct. Nothing he really can "prove." And his decision may not be a popular one. Campus events (rock concerts, football games, etc.) involve a lot of fervent fans and entrepeneurs and a lot of money for the school. Denying or causing parts of that event to be altered, may not make that Security Director the most popular man on campus. Assuming it and sticking to it may require personal courage. For example, the Security Director may review the program planned by a visiting band at half-time of a football game with the school's arch rival. If he finds that program may be apt to enflame the home crowd, he may have to insist that that portion of the program be amended or deleted altogether. He'll have opposition, but if he believes it, he should stick to his guns and win the support of the school administration.

Seldom can his small staff police a large school event with its own resources alone. The Security Director may turn to a contract guard agency or to moonlighting local police for additional help.

If local police augment the school's security staff, several principles are worth mentioning:

- There must be an agreement, in advance and in writing, on the pay of local officers: how much, how paid.
- Cash payments to individual officers should be avoided. Pay the local police department and let them pay the individual officers.
- Assign local police to traffic or other periphery duties, generally away from immediate contact with students. Let campus police officers be in the stadium or wherever the event is taking place, observing the students and clearly seen by the students, prepared to intervene if necessary. The students will better receive police they recognize as their own than they will "town cops." There will be less chance of a confrontation.
- The Security Director must have final approving authority on all local police or contract guards to be hired by the school: on their selection, on their assignment, on their dismissal if they are not competent.

Factors, other than the possible need for local police or contract guard augmentation to his staff, which affect the Security Director's planning for a campus event, include the following:

- What type event is it: Is it one likely to have control problems (a seminar on Faulkner differs considerably in its problems from those likely found at a rock concert)?
- Has this event, or others similar to it on campus, had control problems in the past (*e.g.,* history of past confrontations between two schools for major athletic events)?
- Will the crowd primarily be students or "outsiders"?

- What time will the activity begin? End? Often the time of day and its duration will influence the mood of the crowd and other control factors.
- Location? This will affect parking problems, number of access points to be controlled, etc.
- Will admission be charged? This can affect the type crowd and their mood and will affect security of cash receipts.
- What access controls are necessary? Will tickets be sold at the gate or will there be advanced sales? The method used affects crowd control and the necessity to secure gate receipts.
- What school rules will be involved for students attending the event? For participants in the event?
- What is school policy regarding disorderly behavior, drinking, presence of drugs: arrests, ejection, other?

Traffic

Traffic should move smoothly and quickly. Above all, it must move safely. Roadways must be kept clear of obstructions; be properly lit for night events; have satisfactory directional and warning signs.

Ambulance and fire support should be readily available. Also a tow truck to assist in accidents or to remove stalled or improperly parked vehicles.

Parking lots must be clearly designated and marked, properly lit, patrolled. A Traffic Officer should be available at each lot. If students are to fill this role, they should wear some distinctive identification marking them as members of the security staff. They should be selected for their maturity, judgment, and ability to deal with people. They must be well briefed before the event and well supervised during it.

Pedestrian traffic is controlled in a similar manner: have clearly identifiable officers at key spots to keep crowds moving; be certain that adequate directional and informational signs are posted and visible. Provide officers with some means of communication (walkie-talkies) with their supervisors. Be certain that everyone is well briefed before the event and that supervisors move among them during it.

Dances

Main areas of concern for school dances include the following:

- Control of access to the dance to deter outsiders or gate crashers yet offer easy access to authorized guests.

- Preventing overcrowding, in the overall facility and at particular parts of it (dance areas, sanitary facilities, refreshment areas), to avoid impatient confrontations and safety hazards.
- Fire hazards: blocked exits, blocked stairwells, blocked fire extinguishing systems, smoking in unauthorized areas, improper electrical connections, malicious pranks, etc.
- Presence and use of unauthorized alcoholic beverages; excessive drinking by individuals. If they are drivers and become involved in accidents following the dance, the school may be legally vulnerable.
- Presence and use of illegal drugs.
- Personal arguments and confrontations.

Wherever possible, the security staff, as well as the senior school representative responsible for the overall event, should insist that a specific individual be responsible for the conduct of participants (leader of the rock group, leader of the student body element, etc.). If there is a problem the school official can work through that individual in dealing with all participants.

Rock Concerts

Major problems at rock concerts or similar events are the huge and often unruly crowds they draw; parking; inadequate sanitary facilities; presence of drugs or alcoholic beverages and their effects upon users; controlling outsiders or gate crashers.

The school sponsoring a rock concert will find the cost of insurance very high, not only in premiums to be paid, but also in the stringent public safety standards demanded by the insurance company. The school had best meet those requirements, however, as a minimum, to protect itself against law suits which seem endemic to this type performance.

Partly to meet those insurance requirements, but more importantly to enhance public safety, the Security Director should persuade school authorities to insist upon a detailed, strong, clearly worded contract between the sponsor and the performers. Details that well might be part of such a contract include the following:

- Type performance to include a script or, at least, an outline of its major elements.
- Designation of an individual to be responsible for the entire performing group, on stage and off.
- Number and names of performers.

- Equipment involved; who provides; who installs; any special requirements, and who is responsible for providing them.
- Time the performance is scheduled to begin, duration.
- Encore procedure (if appropriate).
- Statement regarding previous audience reaction to this group's performance in other schools.
- Any special seating requirements.
- Control of backstage area and activities before, during, and after the performance.
- Responsibility for cleaning up particular areas (if appropriate) after the performance.
- Use of complimentary passes: how many, how issued, sample copies, problem referrals, etc.
- If the performer must cancel the performance, it must be done as early as possible and the Security Director notified immediately of that cancellation.
- Rehearsals must be finished no later than two (2) hours before the scheduled performance.
- The performance must start on time; penalities if it does not.
- Back-up performers' availability in emergency.
- The Security Director can terminate the performance at any time he feels a public safety problem has developed.
- The Security Director can use as many security personnel as he feels necessary to insure the safety of audience and performers.

Football Games

This section applies to all athletic events, to one degree or another, but primarily to football because it is the big crowd drawer on any campus and, aside from rock concerts, perhaps also is the biggest drawer of problems.

For football games the Security Director almost always will require outside help to augment his staff. In these cases, whether he uses contract guards or moonlighting local police officers, there should be a clear understanding with the department or firm concerned on the following:

- Selection of individuals (mature, dependable, reliable, good working with people, etc). The Security Director has final approval on all personnel.
- Supervision of contracted individuals (usually best done by his own staff rather than by contracted supervisors).

- Assignment of individuals (usually best to give contracted personnel peripheral assignments not involving prolonged close contact with students; use campus security staff for that).
- Additional equipment required by furnishing agency/firm; cost.
- Cost of personnel, to include overtime and standby costs; how paid and to whom.

He also may have to make similar arrangements for emergency services back-up: fire, medical, vehicle towing.

He must insure that host and visitors (spectators, teams, bands, cheerleaders, etc.) are kept physically separated as much as possible. This is done by having separate entrances, separate dressing rooms, separate seating, and separate sanitary and food service facilities. It doesn't always work out as neatly as he'd like, but if they are separated, as much as possible, particularly in an event between fiercely competing teams, he has a far better chance of avoiding major confrontations.

Reserve seat areas (band, teams, boosters, students) should be agreed upon in advance and be enforced through good signage, barriers, informed Security Officers and ushers.

There should be separate entrance gates for students and general admission ticket holders; special gates for those with complimentary passes or press passes. Throughout the stadium there should be good signage directing pedestrians to these access gates.

There should be good directional signs indicating parking lots, public transportation, medical facilities, seating areas, press booth, concessions areas, rest room facilities, public address facilities, team areas, etc.

There should be separate locker rooms for home and visiting teams. Locker rooms should be clean, well-lit, properly equipped, stocked with necessary supplies, have a means of communication, and be secured at all times.

Officials should have separate locker facilities, similarly equipped.

Ticket booths should be well-lit, clean, and contain telephones. Nearby signs should indicate: ticket prices (when appropriate), gates and portals. Booths should be checked before the game to be sure they contain all the supplies their occupants will need during the game. Resupply, with all the delay it entails, won't be necessary.

Concession stands should be clean, well-lit, adequately stocked, and supplied with all the equipment and material concessionaires will need to do their jobs. They should be equipped with telephones, fire extinguishers, trash cans (ample size, in sufficient quantity, clearly identifiable, properly placed, with appropriate signs). Conspicuous signs should indicate menus, prices, directions to other facilities (if appropriate).

Rest rooms should be clean, well-lit, well supplied with necessary items (nothing is quite so apt to spoil one's day than the absence of a necessary item discovered too late). Rest rooms, like concession areas, must be patrolled by Security Officers throughout the event.

The press box must be properly lit and heated, contain necessary communications equipment (which works properly) and necessary supplies, (*e.g.,* programs) for press representatives.

If it can be done, perhaps with their football tickets, it is helpful for the host school to provide guests with advance information on the host stadium and arrangements for the game (sketch map, routes, parking areas, gates and portals to be used, other stadium facilities).

Score board operators and timers should be mature individuals who work well under pressure. If students are used, they should have an adult supervisor or security representative present with them at all times should arguments develop.

First aid facilities, along with established procedures designating who can use them and how they are to be used, are always necessary.

Crowd control functions of the stadium's public address system must be clarified, put into writing, rehearsed. The Security Director should insure that the public address system is loud enough and clear enough to be understood by all. It should be secured against unauthorized use.

Team and rooters' buses should be parked in a secured area. To prevent incidents, Security Officers should be present when they arrive and when they depart. Similar precautions apply to courtesies rendered game officials.

Many of these responsibilities (*e.g.,* assuring that rest rooms are properly supplied) seem far afield from a Security Director's "job description." They are. They represent likely areas of irritation and conflict by spectators and participants in school events, however, and if he can correct deficiencies before participants encounter them, his job will be made a lot easier.

Football games and other major campus events are meant to be fun. They can be, but only if the Security Director has done his home work long before the event takes place, then supervised his security control plan during the event.

If all goes well, chances are he'll never hear about its going well. He'll know, however, and for the professional that must be sufficient. That is what his job is all about.

CHAPTER 14

FIRE SAFETY

IN A high school gymnasium, scuffling students shove plywood lockers against an electric baseboard heater. That weekend, heat ignites the lockers causing extensive damage.

In a small college, a smoke bomb, thrown in a student dorm as a prank, ignites bedding and causes $10,000 damage before it can be extinguished.

A smoky fire caused by a carelessly discarded cigarette in a Cambridge, Massachusetts high-rise dorm results in a student's death.

Overloaded electrical circuits, short circuits from faulty electrical appliances, improper use of inflammable liquids, arson; all these and other fire hazards pose real problems for the campus Security Director.

It is estimated that campus fires cost more than $20M each year in damages, not to mention individuals injured or killed. Schools cannot afford any of these losses.

In most schools, whoever is responsible for the school's security program also will inherit responsibility for fire prevention. Depending upon the size of the school and its security staff, the Security Director may have a fire expert to supervise this program for him. In a smaller school, he may be able to delegate authority to a staff member who has some training in fire safety. Overall responsibility, however, won't change; it is his. He needs to know about fire safety.

This chapter is not meant to make anyone an expert on fire safety. It does suggest a number of considerations for the individual who may be charged with that responsibility. For more detailed and specific information about fire fighting, he should consult with the head of his local fire department and consider the many texts devoted exclusively to this subject.

Whoever is responsible for the campus fire safety program (most often the Security Director assisted by a staff member who restricts his work to fire safety), he should report directly to the school's President or to a senior vice-president or similar upper-level management official. He'll need that individual's clout to do his job properly.

He must develop programs to meet all the school's fire safety requirements and coordinate fire safety responsibilities among the school staff. He should have authority to demand correction of fire hazards or, if faced with non-compliance, to shut down the offending activity. If he is not the Security Director, he should be a member of all security-related committees and have access to safety and security planning policy meetings. This man will play a vital role in campus security and safety.

His first task is to survey fire hazards, evaluate existing fire prevention measures, and develop ways to cope with the shortfall. The fire safety survey will be similar to that earlier discussed under security surveys, and he might use the checklist suggested in that discussion as a base for his own fire safety survey, developing other items required by his particular situation.

Where appropriate, he should recommend the purchase and installation of fire detection alarms, extinguishing systems (halon, sprinklers, portable fire extinguishers, etc.), and other fire control devices. For this, he may need the expertise of a local fire safety expert and, as the state of the art in fire safety devices changes rapidly, the technical advice of a reputable dealer.

He must develop and implement a fire safety training program for the school staff and for students. It should teach fire hazards and fire fighting procedures, to include how to report detected fires. Fire safety supervisors and members of fire fighting teams must be taught their specific duties and how to use the equipment available to them.

The school fire safety coordinator must continue the training of fire safety supervisors with whatever degree of personal supervision he can exercise over their handling of their delegated responsibilities.

He must investigate all fires and all threats of fires. Then he must take corrective action to reduce the chance of a recurring fire incident.

He should maintain close liaison with local fire officials and stay abreast of developments in the fire safety field (*e.g.*, fire personnel management, training, technical equipment and procedures).

The school's fire safety coordinator can be greatly helped by a fire safety committee, comprised of not more than 6-7 individuals, hand-picked for their personal interest, motivation, enthusiasm and ability, and formally appointed for a specific term of office by the school's head. Committee members should be from upper level school management but should include recognized graduate and undergraduate student leaders. All representing a cross-section of the school's leadership. Activities most involved with fire safety (housing, maintenance, security, plant services, etc.) should be represented on the committee. Appointed individuals must be willing to work and to commit themselves to attaining a good fire safety program. If they don't have that commitment, get someone else without wasting more time.

Depending upon the size of the school and the complexity of the fire safety problem, the fire safety committee might find it necessary to appoint subsidiary committees for more sensitive areas: laboratories, research facilities, hospitals, dispensaries, maintenance shops, etc.

Administratively, all staff and faculty members should be considered fire safety supervisors within their own areas of activity. Practically, however, the Security Director will be lucky to get the full cooperation of a bare majority of most departments heads or leaders of other activities. Those who are not voluntarily committed to the program must be made to understand that he (and the school's head) hold them personally responsible for the fire safety within their organizational areas of responsibility. If he makes that clear enough, they'll pick up the batons he's handed them. They may not like it, but they'll do it. It's enough.

Department heads or other "fire safety supervisors" should be held accountable for:

- Knowing that if there is a fire in their area or activity, they will be held personally responsible until an investigation indicates they had done everything they could to prevent it.
- Knowing fire hazards in their areas or activities and taking measures to reduce or eliminate them.
- Maintaining all areas and all equipment in a safe operating condition; insisting that subordinates and students practice fire safety.
- Instructing their staffs and students in fire safety matters as they pertain to their area or activity.
- At least once a month conducting a thorough fire inspection and maintaining a written inspection record citing specific corrective action taken to correct all discrepancies. The latter should be reviewed by the campus fire safety coordinator.
- Being certain that he is aware of planned or current changes to his facility's physical arrangement or of program changes or other activities that might affect its fire safety plan or create a new fire hazard; taking necessary measures appropriate to that changed situation.

The school's fire safety plan must be in writing. It must be signed by the head of the school, assigning specific areas of fire safety responsibility to individuals. It must make it clear that the program will be enforced. Once that plan is disseminated, lesser administrators must be required to carry out their own responsibilities in it. If the President means it, and everyone know that, it will be a good plan and it will work. It may not be the best plan, but it will work.

Supervisors' fire inspections are necessary to identify conditions:

- That threaten to cause or intensify a fire.
- That might deter fire detection or hinder fire control.
- That might delay or prevent evacuation of individuals from a fire area.

One fire prevention expert has declared that the three chief causes of fire are:

- Men.
- Women.
- Children.

He certainly is correct that the fire safety inspector should be particularly alert to people actions in inspected areas and to indicators of poor knowledge of fire safety, or poor attitudes toward fire safety, or poor practices of fire safety.

The right of fire safety inspectors to enter all areas and all rooms must not be denied. This includes student rooms. Administrators may wish to soften the requirement by reassuring everyone that fire inspectors are not primarily looking for students' pets or students' pot, but the bottom line must be that all rooms are subject to fire inspections at all times. This should be stated in housing agreements.

More important to any fire safety program, however, than the monthly inspection by the department head or his representative, is the daily (sometimes three times daily) inspection by Security Officers as they make their rounds. They must be able to identify and report fire hazards. Such things as the following:

- Unsafe electrical appliances or machinery.
- Unattended electrical appliances or machinery.
- Unusual odors, especially smoke or gas.
- Non-smoking violations.
- Presence of combustibles.
- Blocked exits, blocked fire doors, blocked aisles, blocked sprinklers or other extinguisher systems.

- Jammed fire doors.
- Missing, inoperative or damaged extinguishers; inappropriate extinguishers (water extinguisher in area where electrical fires most likely).
- Rubbish or debris in waste containers or blocking fire exits etc.
- Sticking doors.
- Malfunctioning locks.

Some campus facilities pose special fire hazards. Laboratories, for example, would be concerned with the following:

- Unauthorized individuals using equipment or chemicals.
- Individuals performing unauthorized experiments.
- Individuals working alone and without emergency communication.
- Danger of long hair, beards, or personal dress causing fires.
- Presence of dangerous chemicals or gases.

Whoever makes fire inspections should informally discuss their results with the individuals immediately responsible for the areas/activities inspected and must cite results in a formal written report. They are of little value unless something is done about them. Discussing the problem with the person responsible for the area or activity is the first step toward getting something done about them. They also must be reviewed by the campus fire safety coordinator. He'll verify the appropriateness of the inspector's recommendations or perhaps suggest other measures. He'll be aware of the fire risks noted. Later, he will follow up to be sure that those corrective actions have been taken.

Fire safety inspection reports should contain the following information:

- Date of inspection; date of report.
- Name, title of inspector.
- Conditions noted:
  - Complete description of any unsatisfactory or questionable fire safety matters.
  - Location (building/room/area within a room if necessary to pin down the location).
  - On the spot corrective action taken.
  - Additional corrective measures recommended with a suspense date for their completion.
- Summary of general facility conditions noted: cleanliness, orderliness, neatness, lights, alarms, extinguishers, fire signs, etc. If everything looks fine, he should say so.

- Date, place, name of local supervisor(s) briefed on results of the inspection.
- Space for the fire safety coordinator to review the report and add his comments.

Fire fighting teams should be organized among the staff and faculty. If students are to be involved, use older students first. Before organizing or using teams, however, their use should be cleared with the school's legal counsel as severe insurance or personnel compensation problems may exist.

In organizing training and using fire fighting teams, the following suggestions are offered:

- Team members must be designated for all shifts the school is in operation, not for the 8-4 shift only.
- Alternate key leaders must be appointed.
- Key leaders should have a simple check-in procedure so the fire safety coordinator can know of any critical absences on a particular day.
- Team rosters must be kept current.
- Specific areas of responsibility must be assigned for each leader and for each team.
- When possible, key leaders should be equipped with emergency communication (walkie-talkie radios on a channel separate from that used by the security or the housekeeping departments).
- Training of all team members should include:
  - Organization and responsibilities of the team.
  - School's fire fighting plan.
  - Alarm systems.
  - Fire notification procedures.
  - Location and use of all types of fire extinguishers and other fire fighting equipment.
  - Controls for alarms, fire suppression systems, electrical systems.
  - First aid.
  - Evacuation procedures.

Training must include the proper actions to take when a fire is detected:

- Under all circumstances, regardless of the size or type fire, call the fire department.
- When reporting a fire, give your name/title/department; describe the fire; direct the fire department to an entrance where you will have a guide posted; stay calm.

- When an alarm sounds, even if it seems to be a false alarm, evacuate the area concerned (building if necessary). If it is a real fire, the evacuation may save lives and it will make fire fighting easier. If it is a false alarm, the evacuation interruption may cause peer pressure upon the perpetrator that will prevent recurring false alarms. When the area is declared safe, however, always cause the occupants to return to it. That too may have an effect on repeat false alarms.
- Secure the area: against any stay-behinds; against anyone attempting to reenter the area before the emergency is over; against any intruders.
- Alert fire teams.
- Close windows, doors, cabinets, closets, fire doors to contain the fire.
- Turn off all electrical equipment and machinery, any gas outlets.
- Secure elevators at the ground level. Personnel being evacuated during a fire should use stairways rather than elevators.
- Check to see that sprinkler gate valves are open.
- Post someone to direct firefighters to the fire area.
- Now fight the fire with extinguishers.

When the fire has been extinguished, supervisors must see that it stays out. They then should see that all extinguishers are recharged, as necessary, and returned to their proper locations.

After every fire or every threat of fire, supervisors concerned should submit a complete report on the incident.

The school community will agree that fire safety is very important. The hard part is to get everyone's cooperation in doing everything possible to prevent fires. It takes time, effort, determination, patience, and the tact of a diplomat for supervisors to insist that their staffs take the actions suggested in this chapter.

It can be done, however, and done well. As with many other security matters, however, it requires that the Security Director use all his good judgment, initiative, and enthusiasm. He also may have to throw in all the "muscle" his office has. Whatever it takes to get campus support for his program.

The rewards he'll gain in a single prevented fire, or a single life saved, are well worth all his efforts.

CHAPTER 15

CAMPUS DRUGS

TEN YEARS AGO a Congressional commission on "Safety and Violence in Elementary and Secondary Schools" estimated that 6% of our high school students had used heroin, 8% speed, 5% cocaine, 7% barbiturates, and that marijuana was everywhere. Ten years ago; can anyone doubt its growth since then?

Drugs are part of the campus scene: marijuana, amphetamines (pep pills), barbiturates (downers), tranquilizers, hallucinogens (LSD), heroin, cocaine, morphine, methane phetamine ("speed"), crack. It's there, in one form or another. All the way from elementary schools to the universities.

It is used by the brightest and the dullest students; by the privileged and by the poor; by the prominent and by the nameless faces in the crowd.

In 1986 an All-American basketball player, on the day he signed a lucrative professional contract, died from using "crack" in his college dormitory.

School administrators can handle the problem in several ways. They can say, "No problem here," and by believing it (or at least asserting it), bury their heads in the sand. Abrogate responsibility, pass the handling of drugs on their campus to local police.

Or they can go after them. After the drugs, after the pusher. Standing behind the fact that drug abuse is a felony and refusing to tolerate it any more than they would tolerate any other crime on their campus. Cooperating with federal, state, and local law enforcement agencies but also working to clean their own houses. Driving drugs from dorms, classrooms, campus hangouts.

The latter option is better by far. Toleration can only lead to worse problems. And school administrators have a legal and moral responsibility to provide a safe and secure environment for the entire school community. That means getting rid of drugs, of drug use, of drug pushers. Their primary target should be the pushers, not the users. Stop the pushers and they'll have gone a long way to stop the users too.

The first step is formally spelling out and publicizing a very clear and very strict school policy toward drug abuse:

- Drugs are illegal; they will not be tolerated.
- The school offers help to users; if they do not accept it, and are caught in drug abuse, they will be punished.
- The school will pursue and prosecute drug pushers.

The next step is to develop an investigative capacity within the campus Security Department. The selection, training, and use of this unit was discussed earlier. It should consist of 2-3 investigators, experienced and possessing good ability to deal with people. They will cooperate fully with federal, state, and local agencies and take full advantage of any resources, including intelligence of drug activities on their campus, that those agencies offer.

They must be carefully supervised, probably by a key member of the Security Director's staff or by the Security Director himself, and their work must be coordinated among a very limited and select group in the school's administration and with the school's legal counsel.

Again, their primary target will be the drug suppliers. If they can dry up the sources, they will accomplish much toward cleaning up the campus.

They may employ a variety of techniques, including the following:

- Surveillance teams.
- Surveillance cameras.
- Closed circuit television surveillance of campus areas.
- Dog searches.
- Security sweeps.
- Physical searches of buildings, work areas, residence halls, lockers, etc.
- Security and safety inspections.
- Informants.
- Undercover operatives.
- Telephone hot lines.
- Openly designated drug task force operations.
- Student counseling centers.
- Drug education programs.

Surveillances

Surveillance teams, consisting of trained campus investigators, man positions where they can observe likely areas to determine the presence of drugs, identify individuals involved, and methods used for drug trafficking.

Individuals selected for surveillance teams must be mature, capable of thinking on their feet and using sound judgment, patient to spend long hours in one position (often without tangible results for their efforts).

Cameras with telescopic lens may be used. If so, the investigators must try for a number of good photos of the same activity and, as soon as possible after the surveillance ends, prepare thorough investigative reports of what was seen and what was photographed. Later, they must cross-reference all photos to that report.

Because of the limited area coverage and lack of detail in pictures typical of closed circuit television, television surveillances will be of limited effectiveness against drug abuse. They also are limited by cameras not being safe in the more remote campus areas where drug abuse is most likely to happen. If they are to be used in less protected areas, they probably would have to be disguised miniature cameras. This can become quite expensive and quite sophisticated, too much so for most schools. If television monitoring does detect drug traffic, however, the monitoring station should have the capability of sending someone to the area concerned for direct observation and apprehension. The entire incident must be fully documented by film and an investigative report.

Dog Searches

Dog searches can be very effective but also are very sensitive issues to the entire school community. If you plan to use them, have good assurance before hand that drug activity is taking place in the area to be searched and that the dogs probably will be successful. Otherwise faculty and students may view the operation with considerable resentment and ridicule.

To be a legal search, dog and man must be trained by accredited schools for this type investigation; the handler must be trained to work with that dog; and both dog and handler must be legally in the area they are searching. It can't be just a fishing expedition; they must have reasonable expectations (from other sources) that they will find drug activity, and they must be authorized by school authorities to conduct the search. Dog search teams should include a second officer, if possible, to assist and witness the entire operation.

Security Sweeps

Security sweeps are operations in which a number of Security Officers (may include "deputized" members of the school community under the control of the security staff) "sweep" a designated area to apprehend or drive from school grounds disruptive or criminal elements.

The sweep may be part of a larger plan (identifying and apprehending any trespassers or other individuals who have no business being on campus) or it may focus on a particular problem (group of troublemakers known to hang out near the school gymnasium; group of students who may be picking up drugs from a particular spot or using a particular campus hangout for drug use).

Sweeps can be very effective. Even if they do not apprehend all the individuals targeted, they can show the school community that the school's administration still is in charge of the campus and intends to remain in charge. If they are done right. Done poorly, sweeps earn ridicule for those same administrators and for the security staff.

To be most effective, certain principles should be kept in mind when planning or conducting a security sweep on campus:

- Choose a lucrative target. Your target may be persistent young intruders, drug pushers, youth gangs, outsiders, unauthorized motor vehicles. By previous intelligence (undercover operations, surveillances, informants, other investigative leads), have very good assurance that an improper activity is taking place, exactly where it is taking place, when, what you are likely to encounter in a sweep operation, and that the results will be worth the efforts you'll have to put into the sweep to make it work well.
- Choose a precise location against which to move in your sweep. It may be some specific part of the campus, a building, a room. It must be small enough to be effectively cordoned by the sweep force and the number of suspects small enough to be handled if all are apprehended.
- Develop a good sweep plan. It must be as simple as possible; able to attain good results; include contingencies that could arise before, during, or after the operation. It will involve school officials, campus security, and local police, as necessary.

- Individuals to be given specific details of the planned operation should be severely restricted in number and should be given only information necessary for their roles in the plan. The basic criteria should be: he is authorized to know; he needs to know to do his job. If either element is lacking, the individual should not be informed of the plan's sensitive details.
- Conduct careful and detailed rehearsals of every phase of the planned operation. Do it again and again until it is done right.
- Timing is critical. If the operation is not properly timed, you will commit disorganized elements into an increasingly confused situation. In a situation like that, you probably can't win. And you must "win" a sweep.

  Timing should involve four distinct operations:

  - Assembling forces.
  - Deploying forces to a start point.
  - Conducting the sweep and apprehending suspects.
  - Removing suspects and withdrawing the sweep force.

- Assemble the sweep force at a secured location near the target area (parking lot, building, athletic field, etc). Conduct a final rehearsal and briefing, as necessary. Be sure that all forces start with a common time, common communications, specific objectives. Do not disclose radio use or codes until this point.
- Initial deployment of the sweep force takes place on orders of the sweep control headquarters (all elements equipped with walkie-talkie radios). Sweep elements must reach their designated start points before the actual sweep is to begin, reorganize and adjust their plans, as necessary, prepare to move forward on signal. They must not alert the subjects of the sweep. Surprise is absolutely critical.
- One or more elements of the sweep force must block escape routes. They will hold designated blocking positions while the actual sweep force drives the suspects toward them. All escape routes (buildings, gates, doors, woods, etc.) must be blocked. If this blocking force attempts to become a maneuvering force, gaps will open in the cordon. Suspects will filter through and the operation will fail.
- Once begun, sweeps move quickly, aggressively, explosively. All elements and all individuals involved must be controlled carefully and well. Don't allow over-reacting against the suspects or deny them their rights. Done well, the operation can be a real shot in the arm for your school community, certainly for the security staff. Done poorly, it can do a similar amount of harm.

- If an emergency arises after the sweep has begun (sweep force, or elements of it, needed elsewhere, etc.), stop the operation before it gets out of hand. Fight again another day.
- Enforce communications discipline. Use only authorized channels, call signs, codes. Limit transmissions. Remember that all transmissions probably will be monitored; that can be embarrassing.
- Closing the cordon and apprehending suspects climaxes the operation.
- Removal of suspects and withdrawal of the sweep force, however, are as important as any other phases of the operation. Done poorly, they can spoil the entire operation. If individuals apprehended are students, they probably will be returned to the school for processing. Others should be processed on the spot. "Outsiders" should be delivered to local police who have accompanied the sweep.
- Unless something very definite happens to those apprehended, the effect of a sweep will be lost. If they cannot be arrested (depends upon the situation, gravity of the offense, etc.), they at least can be evicted from the campus. Sometimes it works well to evict trespassers a considerable distance from the campus (or at least at its farthest boundaries). A long walk back to town does much to discourage second thoughts about repeat trespassing on school property; the longer the walk, the greater the discouragement.
- When withdrawing the sweep force, leave behind some of the security staff or other school staff members to prevent escaped suspects' filtering back into the area. Even then, another sweep probably will be necessary later.
- If the sweep must be repeated, avoid predictable operational patterns: same day of the week, same time of day, same deployment areas, etc.

When authorized by the school's head, all buildings, areas, rooms should be subject to public safety physical searches. This should be part of any resident's housing contract.

When a physical search is made of students' rooms or of individual lockers, etc., two staff members should conduct the search. One well might be of a department other than Security, such as the housing office.

Informants, whether they respond to the school's security department or to local law enforcement agencies, probably are the best source of information on campus drug activities. Whatever they reveal should be used only as a lead for an investigation to develop separate information on the alleged illegal activity. A one-on-one confrontation between a source of information and the subject of an investigation always should be avoided. Use the source to point you in the right direction; protect that source at all times.

Informants work for a variety of reasons, not all of which fall under good citizenship. Some motives will not be popular ones (greed, revenge, personal gain, to satisfy psychological needs, etc.). Regardless of their motives, however, sources which are properly controlled and exploited by experienced investigators are valuable investigative aids. Because most informants seek money or other favors for their information, however, it is not likely that they will provide great help for campus investigation. School security departments don't have resources for that purpose. Local law enforcement agencies do. That is why a cooperative investigation (at least up to this point) is important. How far local police, as well as members of the school's staff and faculty, should be involved in campus investigations is discussed elsewhere in this text.

Telephone hot lines, manned 24 hours a day to receive telephone tips on drug abuse, can be very helpful. The caller's anonymity must be protected and the "system" must establish its credibility with the student body or it will not be used.

Drug Task Force operations also are a valid part of any campus drug prevention program. The Drug Task Force, formally approved by the school's head, should consist of representatives of every element of the school community. Their efforts include sponsoring drug education programs (to include school policy regarding drug abuse on campus, dangers of drug abuse, help for drug abusers who really want it, means by which anyone may anonymously alert campus security to drug problems). Their numbers should be limited (too large a group becomes unwieldy), and members should be carefully selected to insure that everyone involved is committed to ridding the campus of the drug problem. Otherwise, individuals who really do not care, or who may object to any counter-drug program, can only hinder the effort.

Student counseling centers should be available to help students who ask for help with their drug problem. Students' rights of anonymity and protection from punishment for voluntarily participating in the program must be protected. The counseling service must establish and maintain its credibility with faculty and students. Otherwise, they might just as well go out of business because they are not going to be getting any clients from the school community.

Whoever is responsible for the school's drug education program should work carefully with the many public agencies which have needed expertise in the field. Many good movies and professionally trained speakers with good examples of drugs and drug abuse paraphernalia are available through these local agencies. Use them.

Drugs are a part of the campus scene. School administrators however, have a legal and moral obligation to make it as difficult as possible for those who would keep them there. Hopefully, this chapter may have suggested ways those administrators, and their Security Directors, can develop effective school drug prevention programs.

CHAPTER 16

PUBLIC RELATIONS

THE Security Officer at the campus gate, in his police-type uniform with badge and distinctive shoulder patch, may be the first staff member persons entering the campus see in the morning, the last they see when they leave.

That Security Officer, by his appearance, demeanor, and professionalism in any personal contact with the school community, represents the school. For better or worse he represents the school. And the school is blessed or tarred with the impression he makes.

Throughout the day Security Officers are in contact with the staff and faculty, students, visitors to the school. They preserve order, enforce regulations, answer questions (lots of questions), provide emergency assistance of all types, and generally are expected to cheerfully pitch in and help in any way they can.

Police relations with their local community are important any time but are particularly sensitive on a school campus. Faculty members and students historically are especially concerned about the administration of law and order and the preservation of what they see as their inalienable rights and freedoms. They will be suspicious of their school's security staff until an individual Security Officer proves that he is competent, that he will listen, that he cares.

Fostering and maintaining good public relations comprise a very important part of the duties and responsibilities of security staff members.

The campus Security Director must train his staff in their public relations role.

Simple things, one would think too basic to mention, must be mentioned. Again and Again. In training and on the job:

- Don't argue with anyone; be polite but firm.
- Treat everyone with dignity and respect.

- Be self-disciplined: avoid anger, sarcasm, profanity, vulgarity.
- Always keep your cool.
- Use "sir," "Miss," "Ma'am."
- Look directly at the person you are addressing.
- Listen; listen carefully.
- Smile.
- Courtesy is contagious: know it, believe it, practice it.
- Think. If you're not sure, call your supervisor.

The training must range from the way a Security Officer should handle all facets of a routine day's duties to approaches he should use in situations which could become very sensitive. Training of Security Officers was discussed earlier. We might now consider special situations they might encounter that could really affect the school's public relations image.

Handling Juveniles

There is a place for children on campuses, and they have a right to enjoy themselves. They are children, yet to learn the full measure of adult responsibilities. They do not have the right, however, to harm others or to spoil others' enjoyment. If they pose a safety hazard to themselves or others, if they are violating school regulations, if their action is offensive or seriously distracting to others, they must be corrected.

In handling juveniles the Security Officer should ask their adult escort to make them behave. If the Security Officer must correct them himself, he should be polite, gentle, but firm. He should never embarrass a child before his peers. No one likes that, even at an early age.

In these days, where children at the earliest ages know the meaning of "child abuse" and can dial "911" on their own, the Security Officer must avoid touching the child and must be cautious in all he says and does to that child. If an incident occurs, he should file a complete report of it in his shift report (to include names of witnesses) if he feels there may be any repercussions to his action.

Handling Mentally Disturbed

A person may be presumed to be mentally disturbed, temporarily at least, when his conduct is so irregular as to be a danger to himself and to others. He may be depressed, angry, frustrated, or feel persecuted. And he is seeking an outlet for those emotions.

The Security Officer has three responsibilities: to protect the public, to protect himself, to protect the individual involved.

He must not treat the individual as if he were not there. He *is* there, and he must be dealt with. Speaking of him to another as if he were not there, or referring to him in the third person, will only add to the individual's frustration and increase chances of a confrontation. The disturbed individual must not be made to feel persecuted or abused. He should be kept calm. He is sick, temporarily perhaps, but sick. The Security Officer should approach him calmly, patiently, gently, reassuringly. He should be as dignified as possible and must convince the disturbed person of his sincerity. He really wants to help. That is the message he must get across.

If restraint is necessary, the Security Officer should try to get help before attempting it. When a person is emotionally disturbed, his body gathers extra strength to meet the emergency it senses. The Security Officer doesn't want to tackle that extra strength alone. Forceful confrontations also are dangerous to the disturbed individual and to any spectators nearby. If he must use restraints, he should get help. The individual probably will be taken to a medical facility off-campus; for this reason, and because local police are more accustomed to this type situation, the Security Officer should first contact his security supervisor who in turn will notify the local police and involve them in the situation.

Handling Intoxicated Individuals

Security Officers often will encounter intoxicated individuals, particularly during campus special events such as football games, rock concerts, etc.

Most of the time the sight of the police uniform and a firm but friendly and understanding approach by the Security Officer will persuade the individual to calm down or to leave the campus. If the individual refuses to leave, or becomes abusive, constitutes a real hazard to himself or others, the Security Officer should notify his security supervisor then follow the supervisor's guidance. Normally, at this point local police are introduced to the situation.

Security Officers must be particularly alert to individuals who are driving while intoxicated. They not only are a great hazard to themselves and to others, but, if the drink has been consumed on school premises, or if school involvement in his mental and physical state can be documented, the school might be held pecuniarily liable for any incident in which that individual is involved. He who sells or gives the drink, in New York for example, can be held liable in a court of law for the effects of that drink.

If the individual is an employee of the school, the Security Officer should quiet him, separate him from others, notify his supervisor.

Concerning student drinking, if the campus security staff can police its own problems, they should do so and not involve town police. Friendly, understanding but firm help "within the family" would be preferred by everyone involved. There is a limit to it, however, and the Security Officer should report all such incidents to his security supervisor so that individual can evaluate who else should be informed of the individual's actions.

Disorderly Persons

If a person becomes disorderly on campus, or shows signs of becoming disorderly (as when his team has lost the game or his ticket turns out to be fraudulent and someone else has his seat), the Security Officer should intervene in a firm but friendly way. While talking with the individual, the Security Officer should assume a relaxed but guarded stance, a safe distance from the individual.

If he knows the individual by name, the Security Officer should address him by that name. It's a disarming technique that works well to cool anger.

If he does not know the individual by name, he might use "Sir," or "Buddy," or some similar address to establish rapport and relax tensions.

Again, if the disorderly conduct continues and the individual refuses to leave or to desist, or if a confrontation seems evident, the Security Officer should contact his security supervisor and follow that individual's directions.

In all instances of aggravated disorderliness, the Security Officer should report details fully in his shift report.

Handling Student Demonstrations

School policy regarding student demonstrations should be written and disseminated throughout the school. It should be clear cut and specific. It should include the role (responsibilities and authority) of the security staff and that of the demonstrators (how to obtain permission to demonstrate, numbers involved, locations permitted, method of demonstrating, etc.). Anything less will make the security staff's job much more difficult and can lead to many problems for the school administrator.

Individuals have the right to peaceably assemble, to free speech, to protest what they consider to be abridgements of their freedoms. They do not have the right, however, to exercise those freedoms at the expense of others. They cannot interfere with the rights of others or place others in danger by their own acts. They have no right to damage school property or the private property of others. The Security Officer facing a student demonstration must walk an often thin line between school authority and the preservation of order and the students' rights to protest.

The Security Officer must remain firm, polite, and calm despite any provocations. He should be on guard against any "baiting" tactics by the disorderly group: name-calling, obscene gestures, racial slurs. If he is part of a solid rank of Security Officers, he should be disciplined not to break those ranks when facing hurled objects or personal insults. So long as the Security Officers present a solid, unbroken front, they can remain in charge of the situation. If it breaks down into man-to-man contests, they already have lost most of the game.

If violence erupts or seems evident, security supervisors should notify local police and request assistance before it gets out of hand. In all instances, they should use minimum force necessary to restore order.

Relying on local police can create problems. It may solve the immediate situation but demonstrators and spectators may then see the security staff no longer as "one of us" but as "one of them." In this instance, linking the campus security police with local police may not be good. The Security Director can handle the problem with his own staff if possible; if he can't, he should call for help before the situation gets out of hand.

Student demonstrations are not the responsibility of the school's security staff alone. The student affairs supervisors (Dean, Head of Department, President, Principal, etc.) should be involved beside the security staff. These latter individuals should not command the security staff, however, in handling the tactical situation. The security staff restores order; the student affairs staff addresses the cause of the disorder.

When a student demonstration has occurred, a complete report of the incident must be prepared for review by the school's head and for the Security Department's records.

Handling Public Media

No Security Officer should speak for his department or discuss any school matters with the public media. If he is questioned about a policy, about an incident, about any security matter, he should refer the questioner to his supervisor.

In screening press media at a school-related event, he should admit only those with honored press credentials (which he should require and inspect). If he has doubts, he should call his supervisor.

The school should have an established policy for dealing with the public media. It should include the designation of a sole press representative for the school administration.

The Security Department should have its own lone media representative, to work with the school media representative or, if authorized, to deal directly with reporters.

The school's security plan should contain an annex spelling out the department's handling of news media matters.

Most often the Security Director will represent the department with the news media. In these duties he might consider the following suggestions:

- Always remember that the media is in business to make money.
- Be aware that "off the record" comments are not always so honored.
- The media has no interest in making you or your department heroes unless it sells their product.
- When a reporter wants information, expect his best cooperation; when you want news space, expect far less. You may have to settle for it.
- Be truthful. One intentional or unintentional false statement can destroy your credibility forever. Demonstrate integrity.
- Learn which reporters know the meaning of personal integrity; be wary of others.
- Always try to have a department spokesman (if authorized) available to meet with media representatives.
- Your cooperating with news reporters does not obligate them to return the favor.
- If an interview or a news report does not go well, it may be best to cut your losses. Public squabbles with the news media usually don't work out well for the Security Director, for his department, for the school. Take your lumps and forget it, but learn by it too.

These suggestions are not meant to suggest that all representatives of news media lack professionalism, integrity, or dependability. It always pays, however, to be on guard. Run an honest department, be truthful and honest with everyone with whom you deal. Then let the chips fall where they will.

CHAPTER 17

HIGH SCHOOLS

SECURITY and public safety problems are greater on a high school campus than at a university, college, or community college.

Locations are a factor. Many secondary schools are located in crowded urban environments where street crime is a fact of life and easily carries over from street to campus.

Age is a factor too. High school students are younger, less mature. Their minds and bodies are groping with change — change within themselves and within their environments. And they are more apt to be influenced by their peers than would their older university counterparts.

Teachers are a factor too. In secondary schools they are apt to be more militant in seeking better working conditions. Understandable, but misdirected this can contribute to school unrest. And many teachers, forced to cope with increased crime, violence, and misbehavior in high school classrooms, abandon their responsibilities to demand good citizenship. From personal fear, lack of interest, incompetency, or weariness in fighting the battle of the hallways, they may simply ignore student misbehavior. What is not seen need not be corrected. Each teacher or other school staff member, who avoids that responsibility as an educator, contributes to the school's security and safety problems.

Compounding the problem, only the largest high school may have a professional Security Director on their staffs, let alone a formal security department.

A harried high school Principal, concerned with more than a platefull of other problems, can't evade responsibility for his school's security. It is his. He might delegate authority to a Vice Principal or to a senior teacher to handle that duty, but he cannot delegate overall responsibility for it. It is his alone.

He can and should charge an assistant to help him meet that responsibility, however, and give him authority to do the job. This chapter offers suggestions for that individual.

Crime, violence, vandalism, and other destructive practices are common to our secondary schools (primary and middle schools too). That's not surprising. Most children emulate their parents.

A sign on an interstate highway in Indiana (heartland of America) exhorted: "God, Guns, Guts Made America Great. Let's Keep All Three." No room there for a reasoned response to a real or imagined threatening situation.

In Pennsylvania, a merchant's billboard proclaimed: "Merry Christmas Everybody. Shotguns, Revolvers, Knives for Sales." So much for the Prince of Peace.

Vandalism and graffiti have become as American as mom, apple pie, football, and McDonald's. In 1975 burglaries, arson, and other destructive acts cost secondary schools an estimated $600 million. Ten years later the tab would be much higher. And taxpayers, tired of the high cost of providing educations for our youngsters, are rebelling more each year against their tax assessments. The school administrator who can reverse that trend in his school, while providing a safer campus where young people can learn, will be a hero first class and will find himself with more resources to devote to his primary task: education.

Experience indicates that:

- The larger the school, the more likely it is to be disrupted by behavioral problems.
- The greater the racial integration in the school, the more likely it is to be disrupted by behavioral problems.
- The greater the racial integration, with white teachers in the school, the more likely it is to be disrupted by behavioral problems.

All schools have disciplinary problems. The causes are not too surprising:

- Breakdown of family ties and family teaching of ethical and moral values.
- Glamorization of violence by television and films.
- Increasing street crimes spilling upon campuses.
- Drug abuse and drug abusers' need for more money to support their drug habits.
- Perceived dislikes for school or neighborhoods.
- Self-serving attitudes reflected in a lack of concern for others.

- Gang or other subgroup pressures.
- Poor role models by parents, teachers, athletes, politicians, doctors, lawyers, etc.
- Our nation's own groping for personal and public ethics.

Whatever the causes, crime, violence, and destruction of school property can ruin a school.

School officials' "solutions" range from a "hands-off" attitude (literally surrendering their schools to disruptive elements) to "get tough" policies that may feature armed police officers in every hallway. The median ground includes use of well placed technical security equipment monitored by a limited security staff, augmented by considerable teacher/student involvement, and all directed by one person.

Key to any high school security and public safety program are the skill, competence, and dedication of the school's Principal. He probably will not be formally trained in means to secure his school against intrusion, violence, or other harm. But secure it he must. The following measures will help:

- Strong and innovative leadership, shared with selected members of the school community, in developing a school security program.
- Appointing a single person to head the school's security program.
- Involvement by the entire school community.
- Regular security surveys and inspections, followed by corrective action.
- Security and public safety staff orientations, emphasizing individuals' responsibilities.
- Establishing and providing for an appropriate level of security for all school and concessionaire property.
- Securing the protection of staff, faculty, students, and visitors by:
  - Visitor control: one entrance; use of sign-in sheets; signs directing visitors to the main office; visitor escorts; challenging unknown visitors.
  - Staff, faculty, and student photo-identification badges.
  - Staff, faculty, and student parking stickers.
  - Adult supervision of student gathering places, hallways, stairs, and other heavily trafficked areas.
  - Adult supervision of all student events.
- Crime prevention information and education programs for the entire school community.
- Continuing review of the security program.

The Principal must encourage a healthy climate of communication within his school community and provide formal and informal networks for it to work. Everyone must feel that his input on school issues is important, and he must know how to go about being heard. Establishing the formal *means* of communication will be the easy part: set them up and publicize them. Setting the climate for communication, however, is more difficult. The administrator will do it by the force of his own personality, by his enthusiasm, and by his willingness to share and to learn from others. By being highly visible on campus; by talking (and listening) to staff, faculty, and students at every opportunity; by showing personal interest in everything that affects his school, he'll establish that climate of communication. And the school will profit by it.

The Principal must establish firm rules for punctuality and attendance. They must be written, published, disseminated, and enforced. He should be able to determine the whereabouts of everyone on his campus at any given time. His staff and faculty, and certainly his students, must know that they are expected to be at their places of duty at times they are scheduled to be there. No exceptions, no excuses. Supervisors' and teachers' records must accurately reflect attendance. Should an incident occur which requires investigation, and all school employees, students, and visitors can be accounted for, the investigator's job will be made much easier.

The Principal must establish and enforce vehicular and pedestrian traffic rules throughout the campus. If there is excessive crowding by either, he might consider staggering work and class schedules to reduce traffic congestion at critical times.

He must overcome teachers' reluctance to become involved in school security.

Teacher-administrator workshops, focusing upon school security problems and the proper ways to handle them, will help. These workshops should emphasize that administrators and teachers are expected, as part of their job responsibilities, to be fully involved with students and with school affairs.

The Principal constantly must work for a curriculum more interesting, more challenging, and more rewarding to teachers and students. The school, for example, should offer worthwhile vocational programs for students not equipped for nor interested in a college education. The curriculum also might include alternative educational programs (night classes, continuing education, credit for life experience) to accommodate working students or those who otherwise find the usual 8-3 daily schedule unacceptable. In sum, the busier he can keep his students, and the more he can challenge them, the less will be his disciplinary problems.

Additional benefits from a busy school are that its facilities offer more value for the dollar and, because of the activity during evening as well as daytime hours, there is less opportunity for vandalism, damage, or theft.

The good Principal will effectively use all his resources: people, buildings, equipment.

If the school does not have a Security Director, *per se,* the Principal should assign that responsibility, as an additional duty, to one other person. His duties would include:

- Initiating and directing all school security programs.
- Enforcing school security policies.
- Hiring, training, and supervising security staff members.
- Conducting liaison with local law enforcement agencies and with fire, medical, and other state and local public safety agencies.
- Maintaining the school's security records.
- Representing the school on all security matters.
- Directing the school's Crisis Management Center; authorizing requests for local police intervention during crises.
- Serving as a member of all school committees dealing with security and public safety.
- Preparing and implementing the school's security budget.

Assisting the Principal and his deputy for security matters should be an advisory committee (call it the School Security Coordinating Committee). Members might include:

- Principal (chairman; makes final decisions).
- Vice Principal (Security Director); Principal's designee for planning and operational control of the security program.
- Faculty representative.
- Community representative.
- Security Department representative.
- Parents representative.
- Local police representative (Sergeant grade or above).
- Fire, emergency medical representative.
- Head Custodian.

If the Principal is to have a successful school security program, he must win the willing support and voluntary involvement in it of the majority of his teachers.

That usually does not come easy. Concerning campus security, many teachers are prone to shrug their shoulders and say, "Let George do it. He's our Sheriff," or "Not my problem. Go rent a cop."

Worse yet, teachers often contribute to the school's security problems. They may do this through insensitivity. I once saw a boy, in a classroom with his peers, forced to wear a crudely labeled sign: "THIEF." Tough treatment for a fifth grader.

Some teachers will belittle or embarrass a student, leaving him no way out except to strike back in whatever way he can. Often by misbehaving. Then, rather than resolve the problem they've created, they pass the buck by sending him to the Principal's office for punishment.

Teachers, who are not dedicated or not competent, bore their students. Inevitably this too leads to behavioral problems. Why not act up, their charges decide, there's nothing else to do.

Some teachers set poor examples. They do it by militancy in contract negotiations (despite court injunctions) or in picket line actions that break the law and are punctuated by violence.

Others simply are not dedicated to their profession and don't want to get involved — with students or with school security.

If the Principal wants his teachers to support security, by his personal leadership and setting the example, he must convince them that their involvement really matters and that their involvement will be rewarded.

However he does it, the Principal must instill in his teachers a sense of the ethical and legal responsibilities of their jobs.

If the school has a formal security staff, it must offer that staff job security, adequate resources, training, and opportunities for personal recognition and advancement (not necessarily within their own department).

School administrators must set certain policies for that security staff:

- Will Security Officers be proprietary or contracted?
- Will Security Officers be in police type uniforms or civilian attire (uniform blazers and slacks)?
- Will Security Officers be armed or unarmed?
- Will supervision of Security Officers be by the school or by local police?
- Are security staff members to have high or low visibility on campus?

The answers depend upon a particular school's environment (urban/rural, high crime area/low crime area, etc.); budgetary restrictions; type students; composition, concerns, and involvement of the remainder of the school community.

Given the choice, and all else being equal, I'd prefer a proprietary staff (or at least contract guards supervised by a proprietary staff); uniformed in a mix of police type uniforms (traffic control or other sensitive posts) and the blazer/slacks combination; unarmed (except for a particularly sensitive area); working under the school's operational control; and highly visible.

Security Officers' duties might include:

- Exterior patrols of campus roads, buildings, and grounds.
- Interior patrols of buildings (hallways, restrooms, cafeterias, other sensitive areas).
- Traffic control.
- Securing and monitoring parking area.
- Manning campus entrance and exit gates for traffic control and informational purposes.
- Questioning individuals suspected of trespassing.
- Informal counseling and correction for minor infractions.
- Recording contacts with individuals, unusual observations and reported incidents, details of their shifts.
- Reporting safety problems.
- Controlling crowds.
- Providing intelligence of possible security problem areas.
- Teaching (staff, faculty, students): drivers education, safety, law enforcement and judicial systems).
- Assisting staff and faculty members, students, and visitors in whatever way is appropriate and possible.

Senior Security Officers' responsibilities might include the following:

- Securing a specific area of the campus during assigned shifts.
- Supervising Security Officers on shifts.
- Participating in security planning.
- Assigning and/or participating in campus investigations.
- Reviewing investigative and incident reports, daily activity reports.
- Liaison with local law enforcement agencies.
- Training the security staff.
- Insuring adequate security and crowd control for extracurricular events.
- Advising the Security Director on potentially disruptive situations.
- During campus emergencies, operation of Crisis Management Center.

The largest schools, with larger and formally organized security departments also might need a communications specialist (responsible for the department's communications systems) and an electronics specialist (responsible for campus alarm and CCTV security systems).

Most high schools have found it necessary (and workable) to involve students in security councils and on security patrols. Participating students (very carefully selected) might be used to:

- Open lines of communication between the administration and the student body.
- Work with the security staff to educate other students and to encourage their participation in programs for personal safety and the protection of personal and public property.
- Report drug abuse, extortion, and similar security problems among the students and which the Security Director can resolve.
- Educate students in the Security Department's role.
- Stop rumors.
- Provide accurate information for parents on security matters.
- Educate (and encourage) other students to be responsible for themselves.
- Improve the school's self-image.
- Define school security and public safety problems.

Another student involvement program might be operation of a Student Crisis Center. The center operates under the supervision of a faculty member but is staffed by students. It deals only with student related problems: hate literature, graffiti, rumors, youth gang disruptions, etc. By accepting and following up on complaints (*e.g.,* teacher conduct, school rules and procedures, security issues, problems between students or student groups), it becomes another unofficial conduit between students and the administration.

If students are to be used in these security roles, participants might be taught:

- Ways to develop self confidence.
- Ways to develop team skills.
- Techniques of conflict reduction.
- Techniques of working with and coordinating with the security staff and with school administrators.
- How to use security equipment (especially walkie-talkie radios with which they might be working).
- *When* to use that security equipment.

- Ways to develop good relationships with local police.
- Rewards for involvement with the school's security program.

Physical systems to enhance campus security have been cited throughout the text (fences, lighting, alarms, CCTV surveillance, locks; use of glazing, mesh or plexiglass to protect windows, etc.).

Several other possibilities include:

- Radar type devices (electronic monitors emit rays toward sensitized surfaces and, by the rays returned, detect intruders).
- Electronic listening devices (sudden sounds in an alarmed area will trigger a light or an alarm in a control station). They do not monitor conversations, only sudden changes in sound levels.
- SCAN (Silent Communication Alarm Network): a transmitter in the form of a pen, pendant, or watch carried by a teacher to transmit a silent alarm to the monitoring station.

If the school has no security department, it might consider a Vandal Watch Program involving "school sitters" living on campus in mobile homes and alerting local police to vandalism or other criminal activities suspected or in progress.

In this program, the school provides, without cost to the vandal watcher, a hard-top stand for the vandal watcher's mobile home (which he provides), connections for utilities, and additional landscaping to enhance the attractiveness of the mobile home and to make it less obtrusive. Public signs, however, warn of the presence of the Vandal Watch Program. School alarms are wired to the mobile home, to be activated whenever the school is closed. If the resident observes suspicious activity, or if an alarm is triggered, he immediately notifies local police who will respond.

The vandal watcher is not expected to investigate or to interfere in any way when an incident occurs. He is there only to alert local police. He is expected to be at home an alert whenever the school is closed throughout the year. During his vacations he must provide a temporary resident who will assume his duties.

Public Relations

Good public relations are critical to any school, whether it be a richly endowed university or a public high school. Not only for fund raising purposes and support of school activities but also to enroll the public's interest and support in the safety of campus facilities and the school community.

The school's administrator must present the image of a school that is interested, caring, progressive; a place where people work to better themselves, not merely a housing for educational devices.

He can show this in a variety of ways: a clean, neat, attractive campus; buildings that are well-kept and free of graffiti or other displeasing scars; teachers who are competent, dedicated, and personally interested in their students; students who care about their school.

He must encourage and insist upon good citizenship by its students. He can foster this attitude through citizenship awareness training, through people appreciation programs, through student projects (plays, posters, concerts, etc.) emphasizing civic awareness, good citizenship, personal responsibilities for law and order, and community pride.

Good public relations projects involving the local community might include open houses and family back-to-school weeks. Parents and the local community as a whole are invited to visit the school and to join student activities.

Other programs which enhance staff, faculty, and students' regard for their school include:

- Smaller classes requiring greater student participation and fostering a one-on-one teacher/student relationship.
- Smaller class buildings or division of a large school into smaller integral units (schools within a school) to reduce an individual student's feeling lost in the crowd.
- More teachers and teachers' aides to allow more time with individual students.
- More teacher-administrator-student interaction on formal and informal bases.
- Better quality teachers (better selection, better pay and other incentives, better opportunities for teachers to *teach,* better supervision by school administrators, the correction or winnowing out of poor teachers).
- Movement toward open campuses and self-directed days. The school gets the basic teaching done, but it also allows students more choice in how to spend "free time." Students sign a "contract" promising not to be involved in drugs, drinking, smoking, vandalism, or other breaches of school regulations. In return, they may select their own use of hours free of scheduled classes. If the student fails to uphold his end of the bargain, he must attend mandatory study halls or other prescribed studies and his parents are notified. In

schools where this approach has been tried, most report better student attitudes, better attendance, less vandalism, and some improvement in grades (usually after an initial drop).

- Counselling: academic, personal, occupational, therapeutic. Not cursory handshakes, waves, pats on the back from a hail fellow well met former teacher who may be a counselor because he found counselling easier than teaching in a classroom, but real, in-depth, caring, well reasoned sharing of thoughts with young people trying to grow into worthwhile adults.
- Night school alternatives to conventional day schools (extended day schools) so students who must work to contribute to their family's support, or who wish to extend their life-experience in avenues open only during the normal school day, may do so.
- Making the school more vigorous, more challenging, busier by:
  - More stringent total requirements for graduation.
  - Written course standards.
  - More stringent basic course requirements with mandatory selection of electives requiring the student to choose some of his work.
  - More elective courses; more variety in electives.
  - Students having less free time during the school day.
- Field programs offering a greater variety of life-experiences and designed to acquaint the student with a broad band of career (or further education) alternatives.
- Broader range of athletic programs, including mandatory participation in intramural sports for students not involved in varsity athletics.
- Involving students in the school's security program by:
  - Posters.
  - Themes.
  - Assembly programs.
  - Student affairs groups.
  - Student public safety patrols.

Total emphasis in this chapter has been upon the high school Principal. His skill as an administrator, as an educator, and as a leader (of his staff and of the student body) is key to the success of the school's security program. He'll set the example and, if he is good, the rest of the school community will follow his lead. He must possess and demonstrate an abundance of dedication to his profession and to his school. He also must use energy and a lot of hard work and imagination to develop and maintain the excellent security program which he does not have the fiscal resources to afford and probably could not be "bought" anyway. In short, he must be the school's leader, in fact as well as in name.

CHAPTER 18

CRISIS MANAGEMENT

IN A CAMPUS crisis situation normal lines of command and control are not adequate. Natural disasters, mass disorders, bomb threats, or hostage situations are not normal problems. Extraordinary measures are necessary to handle them.

An essential element of a school's security program should be developing and rehearsing measures to handle these emergencies. Hopefully, they may never occur. If they do, however, the Security Director who has prepared for them (or for a similar emergency from which he can adapt resources to meet the new requirement) already will have gone a long way toward restoring order on his campus.

The school's chief executive (President/Principal) must focus his attention on the overall functioning of the school and on the majority of his student body. In a crisis affecting only a portion of the school (an explosion in the school's power plant; student takeover of administrative offices, etc.), he may have to designate a temporary emergency staff to concentrate on that particular problem while he sees to the rest of his school. Or the role may be reversed with his concentration on the particular problem while a deputy sees to the routine operations on the remainder of the school.

If his staff (particularly those directly responsible for school security), have done their homework, authority and responsibility to carry out specific plans and contingency options (already spelled out in the school's crisis management plan) already will have been designated. When the crisis situation arises, all that is needed is that administrator's decision to implement the emergency plan.

The Crisis Management Plan might include the following:

- Plans for developing information/intelligence on developing situations.

- Guidance on indicators that a crisis situation is developing.
- Standing Operating Procedures (SOP) for alerting the crisis management staff, local police, emergency services, etc.
- A time-phased plan for initial deployment of school security resources for all conceivable situations; deployment of subsequent additional resources (local and state law enforcement agencies, etc.).
- Additional emergency equipment designated and available, SOPs for issuing, using, returning to storage.
- Mass arrest SOPs (apprehension, arrest, search, processing, medical aid, transportation, documentation, notification of parents, etc.).
- Public relations SOP (one spokesman for the school).
- SOP for communications.
- Documentation requirements (records and reports).
- Time-phased SOP for withdrawing resources.
- SOP for restoring normal school procedures, schedules, etc.
- After-action reports, analysis, follow-up actions.

Staff members selected for crisis management teams should be particularly well qualified, experienced, mature individuals who are capable of remaining cool under pressure, thinking on their feet, and taking appropriate actions without close supervision or direction. They must have good people skills because, in any school crisis situation, most of the problem will be people related.

A Crisis Management Team might include:

- Senior school official (in charge).
- Security Director.
- Local police representative.
- Emergency services representative.
- School public affairs representative.
- School legal representative.
- School conflict management specialist.
- School engineering/housekeeping representative.

A Crisis Management Center should be designated. It should be separated from the Security Operations Center so as not to interfere with routine security operations which will continue during the temporary crisis. For the same reason, it also should be separated from the school's main office where routine business will continue.

Other desirable characteristics for the Crisis Management Center include:

- A central location, physically secured from work interruption 24 hours a day.
- Adequately sized for 5-6 people to work at a time.
- Separate ventilation, HVAC systems.
- Adequately lit, with an auxiliary power system.
- Private telephone lines.
- Bathing, toilet, sleeping, and cooking facilities.
- Stockpiled food, water, medical supplies; supply of expendable items (batteries, paper, etc.).
- Radio communication on a channel separate from either the school's security channel or its routine administrative channel (vehicle dispatch, housekeeping, etc.) but capable of netting with either and with local police and other emergency services.
- Normal office equipment, including copy equipment.
- Wall charts, maps, blueprints of campus facilities (to include utility maps).
- Access to additional stockpiled security equipment (radios, cameras, television cameras, etc.).
- Adequate stocks of appropriate school report forms.

Formal SOPs for equipping, activating, operating and deactivating the Crisis Management Center must be written, taught to appropriate staff members, and rehearsed.

Formal SOPs for handling specific type emergencies should be similarly handled. Emergencies which should be addressed include:

- Natural disasters (fire, flood, wind storms).
- Manmade disasters (fire, fumes, explosions).
- Rumors.
- Hostage situations.
- Bomb threats.

Some considerations in developing these emergency plans follow.

Rumors

Campuses breed and spread rumors as naturally as most lawn fertilizers spread crabgrass. The rumors come in all sizes, shapes, and tones: a racial fight in the gym with a student killed; a drug bust by campus police; police brutality in breaking up last week's student demonstration; a teacher raped by trespassing town youths; the Principal's prolonged illness diagnosed as a nervous breakdown.

Regardless of the specific rumor, handle all of them pretty much the same way:

- Don't react blindly to it. Report it to the school's administrator. Get the correct facts; all of them. Analyze them. Determine the rumor's effect upon the school and the best way to handle it. Then act.
- If appropriate, open the Crisis Management Center as a rumor control center; use the student operated Student Crisis Center as an adjunct to provide an additional means of disseminating information to the student body.
- Refute the rumor with facts. Tell it exactly as it is.
- Later, if it seems appropriate, among older students analyze the rumor as a class project. The discussion may be helpful in handling the next rumor.

Mass Disruptions

Early warning is critical to properly handling mass disruptions. This alert can come from any member of the school community but is most likely to come from three sources: the more responsible students, teachers having the most contact with students, or Security Officers on their daily patrols. The earlier the school's head becomes aware that there is the possibility of mass disruption of his school, the better he can cope with that possibility and, if it occurs, of controlling it with minimum harm to individuals and to school property.

Mass disruptions have a variety of causes:

- Panic resulting from natural or manmade disasters.
- Social problems within the school.
- Social problems outside the school but overlapping on it.
- Outside extremists who would attack the school for social, racial, or economic causes.

The school administrator, and his Security Director, must be careful not to react blindly or impulsively to the threat of mass disorder. To not play "fox and hound" in which a school official runs about chasing lone students or groups of students because he's been baited into a race he can't win anyway. All he'll win, in that situation, is ridicule. There always is time for a cool moment to analyze a situation before taking action that may be irrevocable.

If a mass disruption occurs, school officials might follow the following suggestions:

- Immediately alert the school's security staff and local police (depending upon the gravity of the situation and the school's relationship with the local police) to stand by; alert Crisis Management Team.
- Activate the Crisis Management Center to deal with that particular emergency.
- As quickly as possible, after the existing situation has been analyzed, establish clear policy for the staffs' guidance in areas where they may be affected by the emergency: use of school or outside security resources, authority to effect arrests, closing or evacuating the school, etc.
- Always use the minimum resources/force necessary.
- Display strong resources as early as possible; remove them as early as possible when the situation calms. Don't feed in strength piecemeal, but make it clear that the mass disorder will not be tolerated and that resources are available to end it by force if reason cannot prevail.
- Use a stabilizing staff counselor or other respected school figure, local police representative, etc. to defuse the situation.
- Stage additional resources in a secured nearby area for rapid deployment to the school if necessary.
- Start with a firm but sympathetic approach but don't compromise the security and order of the school and its community at any time. Do not abdicate the authority of appointed school officials. Establish a clear and direct relationship between the school administration and disruptive groups. Address their needs/problems but do not compromise the authority or the safety of the school community.
- Locate, identify, and isolate each major disruptive element. Keep groups in conflict separated from each other and as isolated as possible from other students. Do not tolerate violence by anyone or by any group. Control those who may become violent. Apprehend, arrest, and remove from the campus those who are or have been violent. Talk with anyone willing to talk. Be truthful. Be accurate in any information you disseminate. Encourage teachers and students to get back to their normal routines. Protect your school.
- Arrest all who break the law; remove them from the campus scene.
- When the situation is defused, but not until then, bring conflicting elements back into contact with each other.
- In a phased withdrawal, as the situation improves, remove extra security resources.

Hostage Situations

Americans are prone to think hostage situations can't happen in their schools. They're wrong. They can happen; they have happened. And lives have been lost before they ended.

If a hostage situation begins, school administrators should immediately notify local police. Ending a hostage situation is a police responsibility. They are trained and equipped to handle it.

As soon as possible, police officials should be briefed on:

- Number, type, identity of hostage takers.
- Hostage takers' location.
- Hostage takers' actions.
- Weapons seen, used.
- Number of hostages, identities, descriptions.
- Condition of hostages.
- Hostage takers' demands; specific instructions.
- Any other communications with hostages or with hostage takers.

Police and other emergency service representatives must have a complete description of the school, with specifics (blueprints, utility maps, etc.) on areas affected by the hostage situation.

Administrators must protect individuals not involved in the hostage situation:

- Isolate affected area.
- Evacuate the school:
  - No general announcement; quiet alert to teachers who will evacuate their classrooms and submit an accurate attendance report with any accounting possible for absentees.
  - After students are evacuated, staff and faculty members not critical to the hostage situation, should be evacuated.

The school's Crisis Management Center should be activated. Besides controlling emergency operations, it can be used as a public information center for notification of parents, answering questions, etc. If local police have assumed operational responsibility for resolving the hostage situation, their public affairs representative should be spokesman for school and local agencies.

Police called to the campus scene assume charge of the hostage situation. The school administrator should approve of each of their actions, in advance if possible, and support them. If he disapproves, he should stop them for a review.

A specific school's involvement with local police to resolve a campus crisis situation of any sort depends very much on the unique relationship between that school and that police department (history of past interactions, the degree of liaison and advanced planning that has taken place between the school's security staff and local police; the competence and experience of the leaders involved; and the personalities and personal relationships among those leaders).

Some basic principles, however, are valid in any school/police department crisis interactions:

a. Police alerts
   - If possible, use pre-arranged codes for secure telephone channels.
   - Change alert codes after each use and periodically to deter unauthorized use of the codes.
   - Have a pre-arranged system by which the police department can quickly verify a coded alert call.
   - Keep the initial alert message brief but with sufficient information for the police to make a good initial assessment of the situation.
b. A responsible school official should be waiting to meet the police department's representative when he arrives on campus.
c. After the alert, the first direct contact by the police with the school may be the police department's liaison representative. As he will make an initial assessment of the crisis situation and determine initial police involvement, he must be quickly but thoroughly briefed by the school administrator or Security Director on the who, what, when, where, why, and how of the situation.
d. The police liaison representative will report to his supervisor who then will dispatch police units to the school. The commander of those initial police units should:
   - Contact the school administrator.
   - Set up a police command post.
   - Require a briefing on the tactical situation.
   - Assess the current situation and report to his police supervisor, requesting additional police or other emergency aid if necessary.
   - Prepare police reinforcements' staging areas, routes, points for police to enter the campus, campus deployment and initial tactical use.
e. Police main force units and other emergency units (fire, medical, bomb disposal) usually will stage away from the campus, reorganize, and deploy to the school on signal from the police official in charge on campus.

f. Tactical control of operations:

- Once police enter a campus crisis situation, the ranking police official present assumes operational control of all security elements (school and local police).
- The ranking police official present on campus is responsible for maintaining contact with the school's administrator.
- Actions by the security force (school and local police) must be approved by the school administrator or his representative. Once he gives approval for an action, he should support it. If he does not approve, it should be stopped and mutually reviewed.

Bomb Threats

Bomb threats are a serious concern to security staffs anywhere but particularly in a school. At the least (a threat but no bomb present) they mean severe disruption of school routine (usually the reason for a false threat); at worst (a real bomb or incendiary device) they can mean great loss of property and lives.

In most cases bomb threats are only threats and will not be carried out. No threat of this type, however, can be ignored. Each must be assessed on its own merits and in light of what has gone before. If the school has not been subjected to explosive or incendiary attacks before, chances are they will not be used now. If they have been used before, using one now will be that much easier, hence more likely.

If the school administrator has evacuated the school, or at least curtailed classes, at previous bomb threats, the chances are very good that from time to time he'll have more of them.

Whether the threat is real or not is one of the decisions he must face. As the bomb threat crisis develops, he may have to face more:

- Whether to notify or not to notify police.
- Whether to search or not to search.
- Whether the bomb or incendiary device has been found or whether it (or others) still remains.
- Whether to evacuate the school, part of the school, or none of the school.
- Whether a detonation has occurred, may yet occur, will not occur.

Central switchboard operators, security personnel manning the Security Operations Center, and department heads or supervisors likely to be targeted for a bomb threat should be oriented on the proper handling of such calls, threat assessing, and emergency procedures to include bomb searches.

Bomb threat checklists, similar in content to the one shown at the end of this chapter (Figure 4), should be available at all times near phones likely to be used for these calls.

If a bomb threat call is received, the person receiving it should take the following actions:

- Keep talking with the caller. Seem interested, friendly, not biased toward him. It may be necessary to pretend that the telephone line is garbled or that the receiver of the call does not understand and must have information repeated—all to keep the caller talking while the listener evaluates the information he is receiving. He must get all the information he can, particularly: Where is the bomb? What does it look like? What kind of bomb is it (incendiary, explosive, etc.)? When is it set to detonate?

  As the call continues, the person receiving it should make notes. It is very important that he recall the exact wording of the caller's basic message as it may locate and identify the explosive device for defusing. It also may indicate whether the threat is real or not. For example, if the caller cannot identify a portion of the school gymnasium where the bomb is allegedly hidden it probably is not there at all. If possible, and without disturbing the caller, the listener should alert another operator or supervisor to quietly listen to the conversation.

- Complete the bomb threat checklist immediately after the call and as thoroughly as possible.

- When the call is completed, the listener should notify his immediate supervisor.

- The listener should not discuss the threatening call with any other individual. Reporters and other outsiders should not be given any information on the bomb threat unless that is done by the officially designated school (or police) public affairs representative.

Reporters and other outsiders should not be allowed on the campus during a bomb threat emergency: only local police, bomb disposal specialists, and other necessary emergency services personnel should be admitted.

If police have not assumed operational control of the bomb crisis situation, the Security Director or, in his absence, the senior security staff member on duty in the Security Operations Center acts as follows:

- Notifies all individuals on the school's official emergency notification list; notifies police and other emergency agencies.
- Directs and supervises a search of all or part of the campus, as appropriate.
- Directs and supervises evacuation of the school if necessary. The decision to evacuate should be made by the senior school official present, with the advice of the Security Director or his representative, not automatically by the security staff.
- Directs and supervises reoccupation of the school after the emergency has ended. Again, the basic decision to reoccupy a threatened school should be made by the senior school official present, on advice of the Security Director.

Following any evacuation of a threatened school facility, if that facility is declared safe for reentry during the normal school day and staff and pupils remain available, it should be reoccupied. This discourages future false alarms.

If the alleged detonation time is imminent (few minutes), and the "bomb's" location is given by the caller, the area concerned should be evacuated until a search is completed. If time permits, careful analysis of the credibility of the threat should be part of an evacuation decision because, if the school is evacuated, whether for a real or a simulated bomb, the perpetrator has accomplished his basic purpose of disrupting the school.

All calls activating bomb threat emergency plans should be as brief as possible to keep lines free and to expedite emergency notification procedures. The line on which the bomb threat was received should be kept free as the individual who made the first call might call again with additional information.

The Security Director should assure that a log of all communications and other happenings is begun as soon as possible and that the log is maintained throughout the emergency.

Use of two-way radios should be discontinued within the threatened areas until the bomb threat is resolved as it could set off an electrically detonated device.

Depending on the time available and other factors, the initial evacuation decision may have to be made by the senior security supervisor present. As school officials reach the scene, however, or police arrive to assume operational control of the crisis situation, responsibility for that decision automatically passes to them. The basic rule, however, is that, if in doubt, evacuate.

Staff members and teachers should take immediately available personal effects with them, leaving cabinets, drawers, and all doors slightly open to minimize blast effects.

If the bomb is located, personnel should not be evacuated through that area (use different exits, different stairwells, etc.). If the school is multi-story, evacuation should be made to at least two (2) floors below the located or suspected bomb location.

Time permitting, one hour as a guide, search of floors cited in the bomb threat could be conducted by school staff search teams under the supervision of the security staff. Staff members, custodians, and teachers should be used with the search teams in areas where they habitually work. They'd be most likely to notice a suspicious or foreign object (*e.g.*, unexplained length of pipe on the floor of a rest room, a carton misplaced in a class room, a package behind a fire extinguisher, etc.).

Bomb searchers should first listen, then divide the area or room into sectors, search one sector at a time, search from the floor up to the ceiling, and notify the Crisis Management Center (if operating; if not, the Security Operations Center) as a room or area is cleared.

Members of bomb search teams should not touch suspicious objects. They should summon help from the Crisis Management Center which will be in immediate contact with law enforcement agencies or Army demolition teams trained in handling explosive devices. All examination, transport, disposition of possible bombs or incendiary devices must be left to agencies trained and equipped in bomb disposal techniques.

The chances of a school's receiving a bomb threat or similar warning are high. In most cases the threat will not materialize into an actual bomb situation. No one, however, can afford to take that chance. The school's administrator, or his Security Director, should have bomb threat emergency plans prepared and rehearsed. With that preparation, even if the threat is legitimate, the risk of harm to individuals or to the facility is greatly lessened.

BOMB THREAT RESPONSE

Your name: _____ Date: _____

Location: _____

Time call received (AM/PM): _____ Number/Extension: _____

Exact words of caller: _____
_____

Person caller requested: _____
_____

I said: _____
_____

WHERE is it located (record exact words)? _____
_____

WHEN will it explode (record exact words)? _____
_____

WHAT does it look like (record exact words)? _____
_____

WHAT kind of bomb is it (incendiary/explosive) (record exact words)? _____
_____

WHO is the caller (record exact words)? _____
_____

      Address? _____

      Organization? _____ Phone number? _____

      Other statements: _____
      _____

Time caller hung up (AM/PM) _____

The caller's:

    Sex: Male _____   Female _____   Age: _____

    Voice: Fast _____ Slow _____ Distinct _____ Disguised _____

    Language: English _____   Foreign (identify if possible) _____
             Educated _____   Simple _____   Profanity _____

    Tone: Loud _____   Soft _____   Harsh _____

Figure 4

Figure 4 *(continued)*

Accent: Local _____ Regional (identify) _____

Manner: Calm _____ Angry _____ Emotional _____
Laughing _____ Crying _____ Deliberate _____

I can _____ cannot _____ imitate unusual characteristics of the caller's
voice: _____

The caller's voice was _____ was not _____ familiar to me: _____
_____

Background noises: _____
_____

Local call _____ Long distance _____

I notified:
School: _____
_____

Police: _____ Fire: _____
Others: _____
_____

Signed: _____ Date: _____ Time (AM/PM): _____

INDEX

195